CW00863379

# *PILOT'S FLIGHT OPERATING INSTRUCTIONS*

*FOR*

# ARMY MODEL P-61C AIRPLANE

This publication contains specific instructions for pilots and should be available for Transition Flying Training as contemplated in AAF Reg. 50-16.

Appendix I of this publication shall not be carried in aircraft on combat missions or when there is a reasonable chance of its falling into the hands of the enemy.

*Published under joint authority of the Commanding General, Army Air Forces, the Chief of the Bureau of Aeronautics, and the Air Council of the United Kingdom.*

*NOTICE: This document contains information affecting the national defense of the United States within the meaning of the Espionage Act, 50 U. S. C., 31 and 32, as amended. Its transmission or the revelation of its contents in any manner to an unauthorized person is prohibited by law.*



FORT WAYNE PRINTING CO.
FORT WAYNE, IND. 10-15-45--4,400

*25 JULY 1945*


ISBN #978-1-4116-8900-8

# POLICY GOVERNING DISTRIBUTION AND USE OF THIS PUBLICATION

**Instructions Applicable to U. S. Army Personnel:**

1. This publication is intended for technical aid and education of military and civilian personnel engaged in promoting the war effort. Its maximum distribution and use is therefore encouraged. However, since the publication is "restricted" within the meaning of AR 380-5, the following security regulations will be observed:

a. *Members of Armed Forces and civilian employees of War Department* will be given access to this publication whenever required to assist in the performance of their official duties (including expansion of their knowledge of AAF equipment, procedures, etc.).

b. *Personnel of War Department contractors and subcontractors* may be given possession of this publication, on a loan basis, or knowledge of its contents, only when required to assist in the performance of War Department contracts. Releases will be made in accordance with the requirements of T. O. No. 00-5-2.

c. *Representatives of other governments* will be given possession of this publication, or knowledge of its contents, only in accordance with AAF Letter No. 45-6.

2. This publication is restricted because the information contained in it is restricted. It does not follow that the physical article to which it relates is also restricted. Classification of the materiel or component must be ascertained independently of the classification of this document.

3. Neither this publication nor information contained herein will be communicated to press or public except through Public Relations channels.

**Instructions Applicable to Navy Personnel:**

Navy Regulations, Article 76, contains the following statements relating to the handling of restricted matter:

"Par. (9) *(a)*. Restricted matter may be disclosed to persons of the Military or Naval Establishments in accordance with special instructions issued by the originator or other competent authority, or in the absence of special instructions, as determined by the local administrative head charged with custody of the subject matter."

"*(b)* Restricted matter may be disclosed to persons of discretion in the Government service when it appears to be in the public interest.

"*(c)* Restricted matter may be disclosed, under special circumstances, to persons not in the Government service when it appears to be in the public interest."

The Bureau of Aeronautics Aviation Circular Letter No. 90-44 contains the following paragraph relative to the use of aeronautical technical publications:

"Par. 8. *Distribution to All Interested Personnel.* In connection with the distribution of aeronautical publications within any activity, it should be borne in mind by the offices responsible for such distribution that technical publications, whether confidential or restricted, are issued for use not only by officer personnel, but also by responsible civilian and enlisted personnel working with or servicing equipment to which the information applies."

**Instructions Applicable to British Personnel:**

FOR OFFICIAL USE ONLY.—Not to be published nor communicated, either directly or indirectly, to the press. Subject to the restrictions currently applicable to official Air Publications.

## LIST OF REVISED PAGES ISSUED

*NOTE:* A heavy black vertical line, to the left of the text on revised pages, indicates the extent of the revision. This line is omitted where more than 50 percent of the page is revised.

## ADDITIONAL COPIES OF THIS PUBLICATION MAY BE OBTAINED AS FOLLOWS:

AAF ACTIVITIES.—In accordance with T. O. No. 00-5-2, base Air Inspectors, Technical will submit requisitions (AAF Form 104B) to:

Commanding General
Fairfield Air Technical Service Command
Patterson Field
Fairfield, Ohio
Attn: Publications Distribution Branch

NAVY ACTIVITIES.—Submit requests to Chief, BuAer, Navy Department, Washington, D. C., Attention: Publications Branch on order form NAVAER-140. For complete listing of available material and details of distribution see Naval Aeronautic Publications Index, NavAer 00-500.

BRITISH ACTIVITIES.—Submit requirements on Form 294A, in duplicate, to the Air Publications and Forms Store, New College, Leadhall Lane, Harrogate, Yorkshire, England.

# TABLE OF CONTENTS

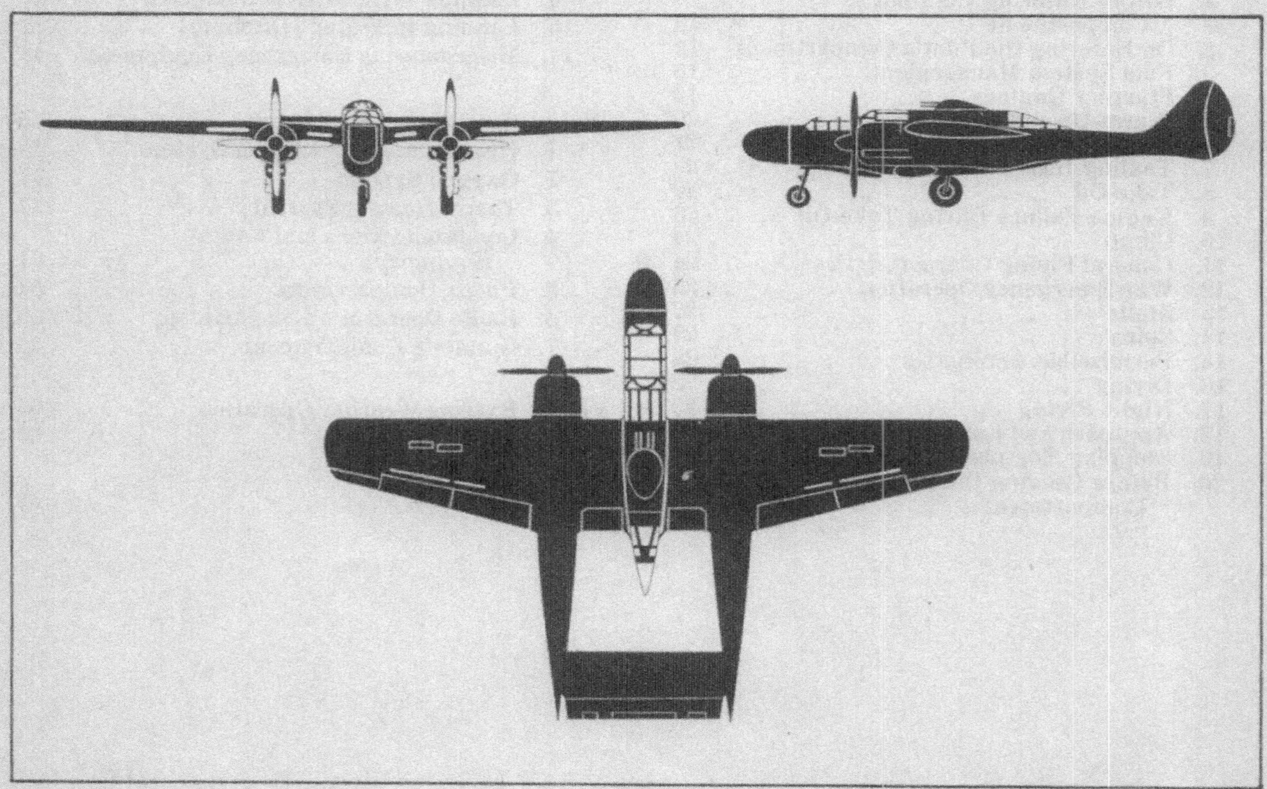

***Figure 1 — P-61C Airplane***

## 1. GENERAL.

The Northrop P-61C night interceptor airplanes are three-place, twin-boom, midwing monoplanes, powered by R-2800-73 turbosupercharged engines. They have a wing span of 66 feet, a length of 49 feet 7 inches, and a height of 14 feet 8 inches. The approximate gross weight, without external load but with full fuel and armament, is 31,500 pounds. The airplanes are armed with four .50 caliber machine guns and four 20-mm cannon. They are equipped with four wing racks to carry fuel tanks, chemical tanks, or bombs. The nacelle has separate compartments to accommodate each member of the crew. The radio operator's compartment is in the aft section of the nacelle and is separated from the pilot's and gunner's compartments by a turret and radio equipment.

## 2. FLIGHT CONTROLS.

*a.* The ailerons, elevator, and rudders are conventionally operated by a control column and rudder pedals.

*b.* RETRACTABLE AILERONS. — Four retractable ailerons, two in each outer wing, are interconnected with the conventional ailerons and function simultaneously with control movement. Important effects of the retractable ailerons on the airplane flight and handling characteristics are:

(1) A high rate of roll can be produced with unusually low control force. The conventional ailerons balance the action of the retractable ailerons and serve to lend "feel" in aileron control.

(2) At low speeds, as in landing, somewhat more deflection of the control wheel is required to produce a given rate of roll than with a conventional aileron system.

(3) There is no tendency for the line of flight to be deflected downward as moderate roll is induced by the use of the retractable ailerons.

*c.* FLAPS.—Full flap deflection is 60 degrees. Flap up-lock position lights are located below the pilot's compartment left rail, just forward of the flap control quadrant. (See 6, figure 7.) The position lights are on only when the flap lock is open, and during the first 10 degrees of flap operation from the full position. The flap control lever (see 5, figure 7) must be moved two-thirds of the way from "NEUTRAL" to "DOWN" to release the flap locks mechanically. During this portion of the arc the flap control lever usually works stiffly. As the lever moves through the last one-third of its arc it releases the hydraulic pressure which moves the flaps. When the flaps are locked in the "UP" position the lever should be returned to "NEUTRAL." The flaps must be locked up at airspeeds in excess of 175 mph IAS.

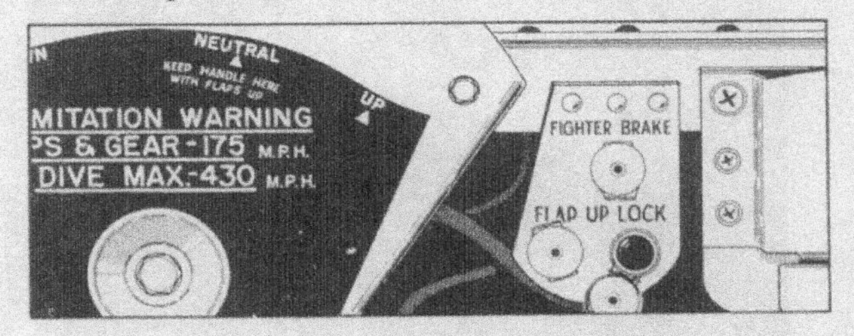


**Figure 2 — Wing Flap and Fighter Brake Indicator Lights**

*d.* TRIM TABS.—The ailerons are not equipped with trim tabs. The rudder tabs are controlled by a dial knob and the elevator tabs by a wheel. (See 10 and 12, figure 7.) Two non-adjustable spring tabs on the elevator reduce control forces, especially at high speeds.

*e.* FIGHTER BRAKES. — Fighter brakes are installed in the outer wings. They consist of extendible panels in the upper and lower surfaces of the wing, similar to dive brakes, and afford a means of rapid deceleration in flight. Fighter brakes cannot be operated unless the landing gear is locked in the "UP" position. An emergency retraction system is provided. (See figure 3.)

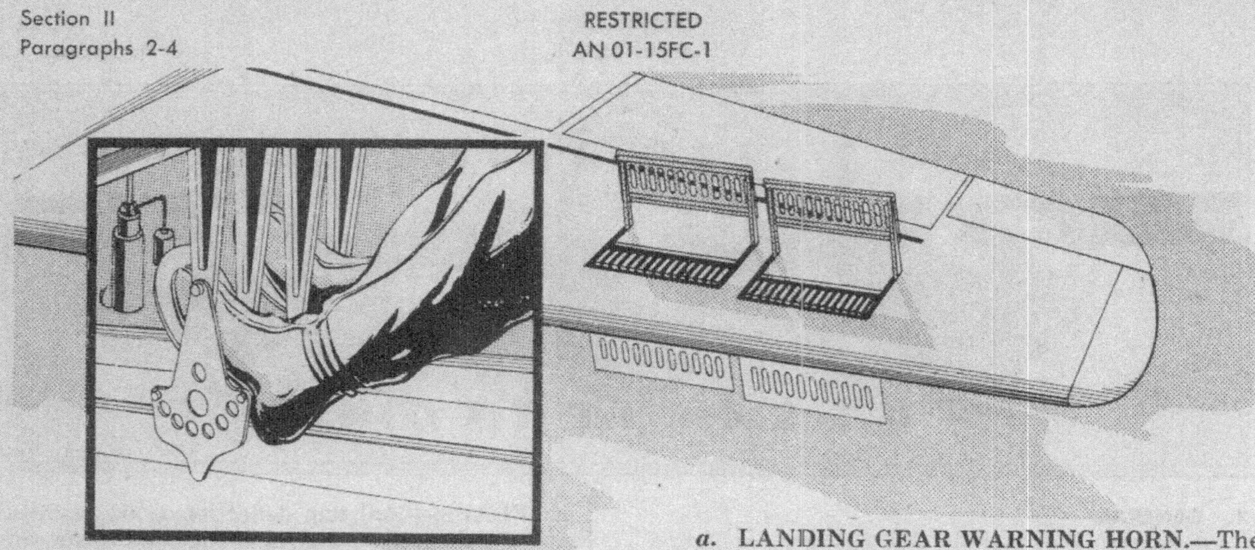

Figure 3 — Fighter Brake Operation

*f.* FLIGHT CONTROL LOCK.—The flight control lock is located to the right and aft of the pilot on the floor of the gunner's compartment. The surface controls are locked in "NEUTRAL" and the throttles in the "CLOSED" position when the flight control lock is engaged. (See figure 4.)

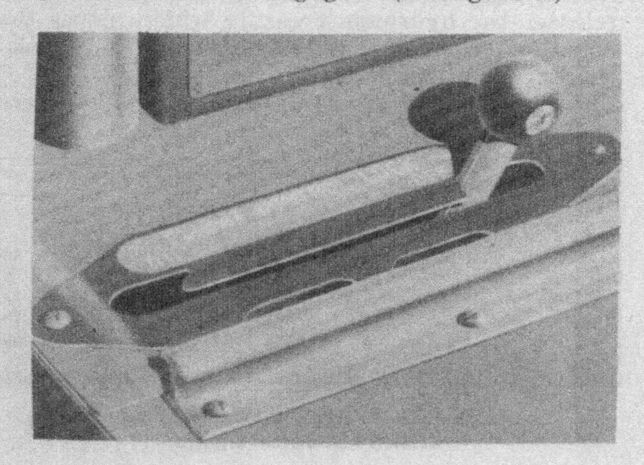

Figure 4 — Flight Control Lock

### 3. LANDING GEAR CONTROLS.

The hydraulically-operated tricycle landing gear retracts aft into the crew and engine nacelles. The control (square knob) is located forward and left of the control column. (See 11, figure 6.) When the gear is locked in the "UP" position return the knob to "NEUTRAL." When the gear is "DOWN" *do not* return the knob to "NEUTRAL." The flap and gear position indicator is in the lower right-hand side of the instrument panel. (See figure 15.) A safety lock is provided to prevent inadvertent retraction of the landing gear before the airplane is airborne.

*a.* LANDING GEAR WARNING HORN.—The warning horn will sound if the gear is down and not locked and either throttle is closed to 15 inches Hg manifold pressure or less. The warning horn release is on the engine control quadrant. (See 14, figure 7.)

*b.* BRAKES.—Pressure for brake operation is supplied from the hydraulic accumulator. An emergency air brake system is provided for use in the event of hydraulic brake failure. (See 16, figure 6.) The parking brake control is between the rudder pedals.

### 4. FUEL SYSTEM CONTROLS.

*a.* Four self-sealing fuel tanks are installed, one in each inner wing and one in each engine nacelle. They have the following capacities:

| | |
|---|---|
| Engine nacelle tanks (2) | 205 US (171 Imp.) gallons each |
| Inner wing tanks (2) | 115 US ( 95 Imp.) gallons each |
| Total | 640 US (534 Imp.) gallons. |

In addition, four pressurized, drop tanks may be installed, one under each inner wing and one under each outer wing. Either 165 US (137 Imperial) gallon drop tanks or 310 US (258 Imperial) gallon drop tanks may be installed. Total maximum fuel capacity is 1880 US (1556 Imperial) gallons. (See figure 5.)

*b.* BOOSTER PUMPS.—Each main fuel tank is equipped with a two-speed, electric, fuel booster pump. The switches for the pumps are located on the pilot's electrical switch panel. (See figure 10.)

*c.* FUEL QUANTITY GAGE.—The gage, mounted on the right side of the instrument panel, consists of four dials and indicates the amount of fuel in each of the four main fuel tanks. (See figure 15.) Fuel quantity gages are not supplied for the drop tanks.

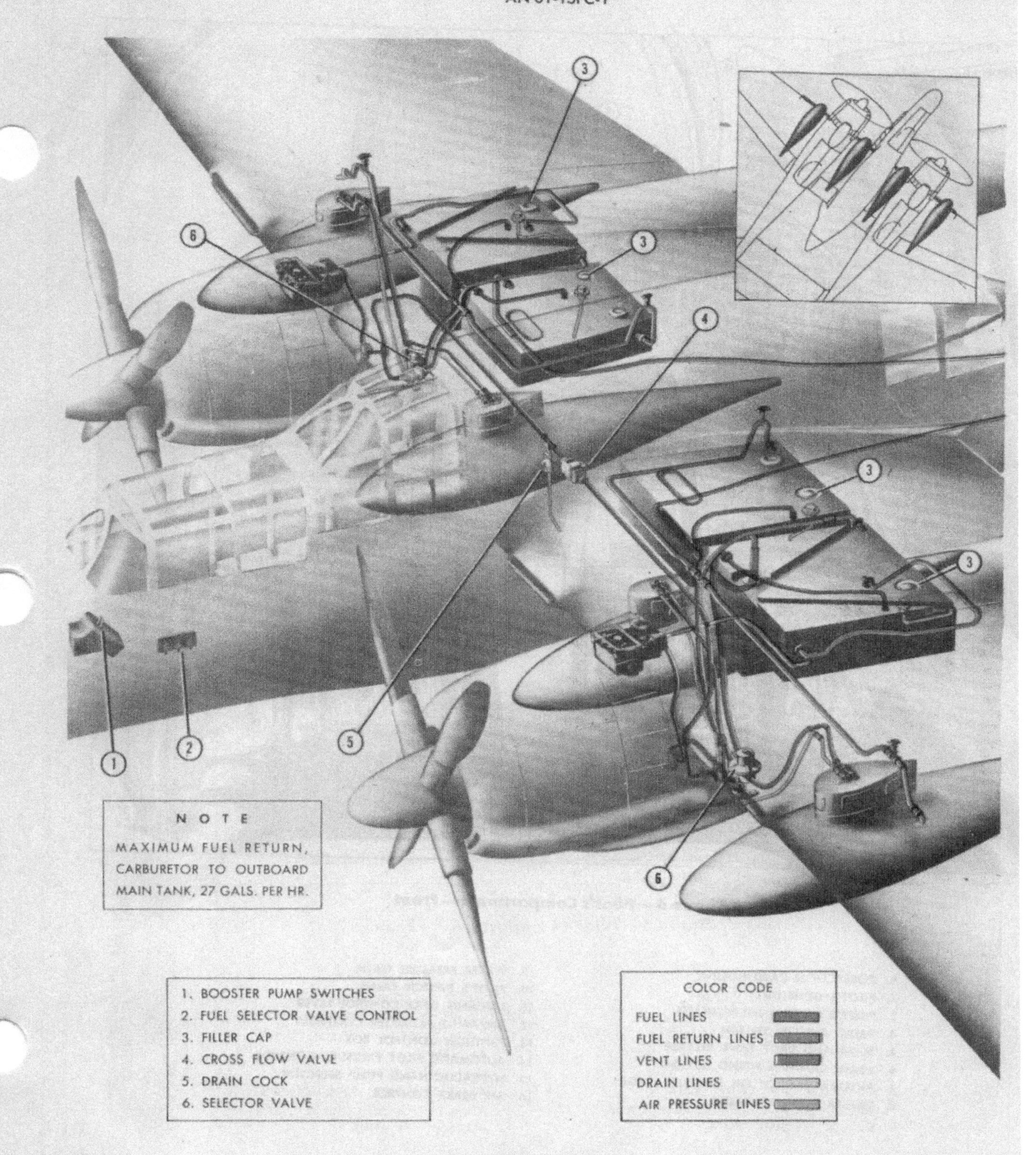


1. BOOSTER PUMP SWITCHES
2. FUEL SELECTOR VALVE CONTROL
3. FILLER CAP
4. CROSS FLOW VALVE
5. DRAIN COCK
6. SELECTOR VALVE

| COLOR CODE |
| --- |
| FUEL LINES |
| FUEL RETURN LINES |
| VENT LINES |
| DRAIN LINES |
| AIR PRESSURE LINES |

***Figure 5 — Fuel System***

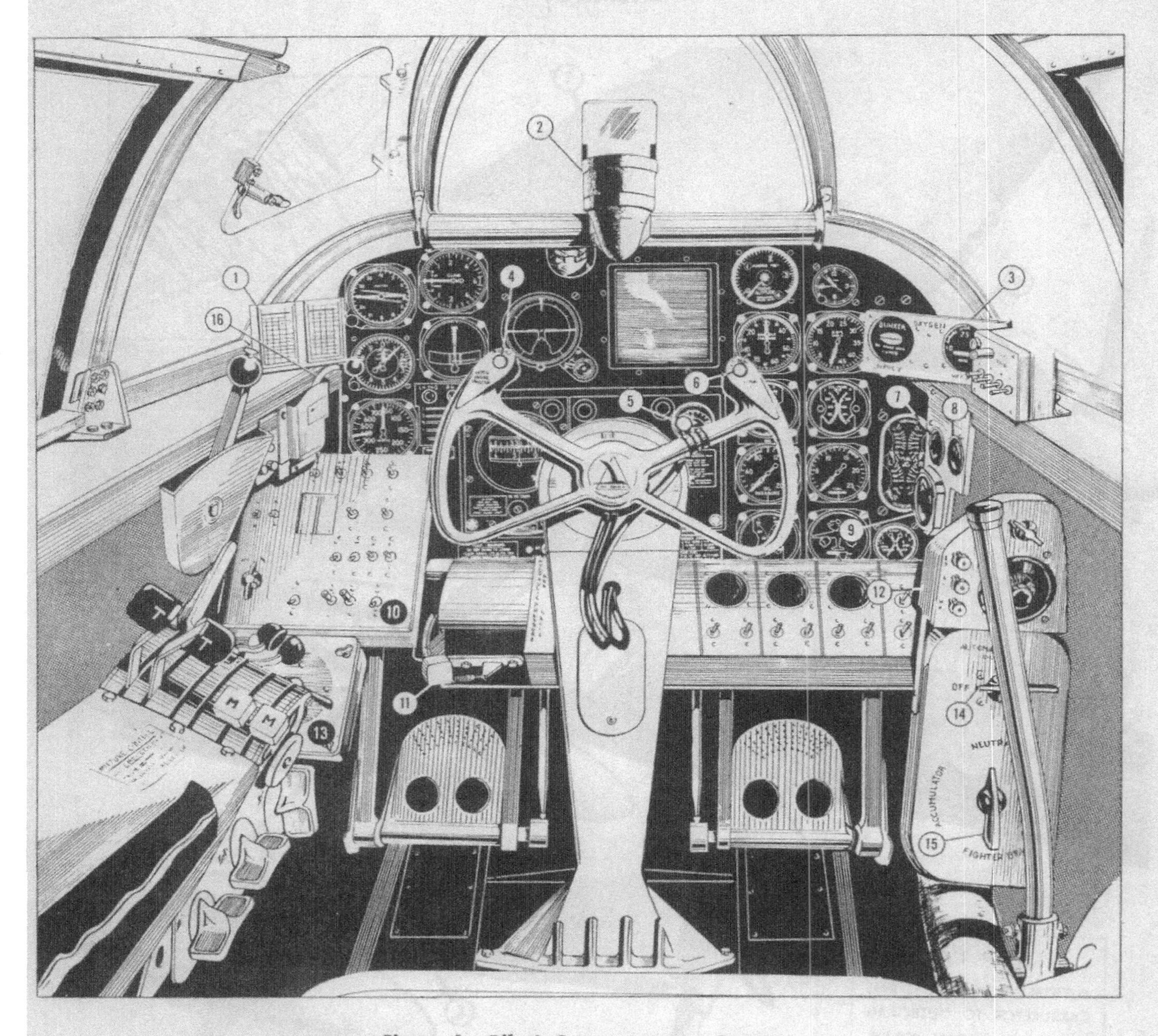


***Figure 6 — Pilot's Compartment — Front***

1. CORRECTION CARD HOLDER
2. PILOT'S GUNSIGHT
3. PILOT'S OXYGEN INSTRUMENTS
4. RADIO BUTTON SWITCH
5. BOMB AND DROP TANK RELEASE
6. 20-MM CANNON FIRING BUTTON
7. AUTOMATIC PILOT OIL PRESSURE GAGE
8. DE-ICER PRESSURE GAGE
9. WATER PRESSURE GAGE
10. PILOT'S SWITCH PANEL
11. LANDING GEAR CONTROL LEVER
12. AN/APN-1 ALTIMETER CONTROLS
13. IGNITION CONTROL BOX
14. AUTOMATIC PILOT PRESSURE CONTROL
15. HYDRAULIC HAND PUMP SELECTOR
16. AIR BRAKE CONTROL

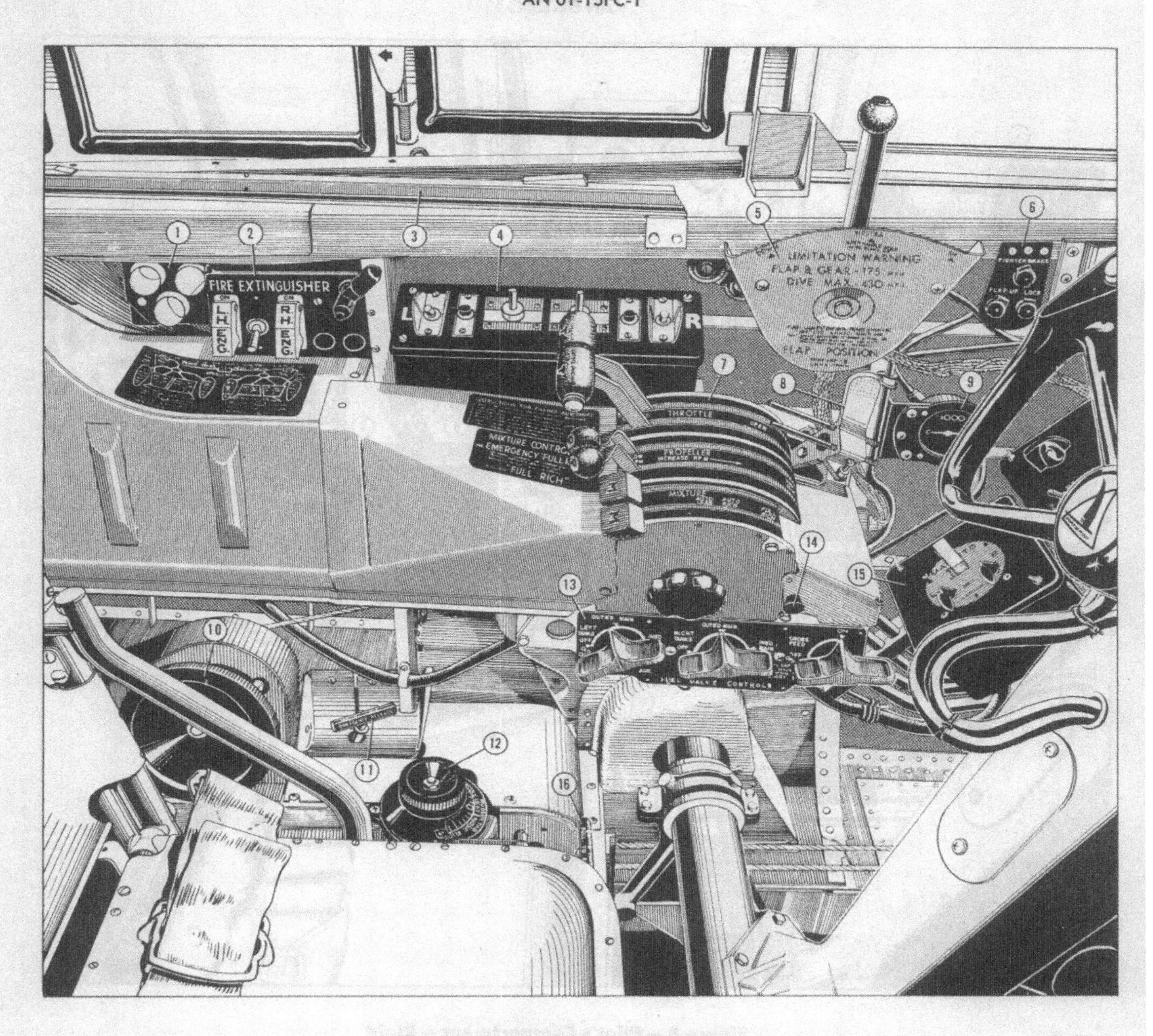


*Figure 7 — Pilot's Compartment — Left*

1. SPARE LAMP BULBS
2. FIRE EXTINGUISHER CONTROL PANEL
3. NIGHT BINOCULAR TRACK
4. PROPELLER CONTROL BOX
5. WING FLAP CONTROL QUADRANT
6. WING FLAP AND FIGHTER BRAKE INDICATOR LIGHTS
7. ENGINE CONTROL QUADRANT
8. WATER INJECTION THUMB SWITCH
9. AIR BRAKE PRESSURE GAGE
10. ELEVATOR TRIM TAB CONTROL
11. LANDING GEAR EMERGENCY RELEASE
12. RUDDER TRIM TAB CONTROL
13. FUEL SELECTOR VALVE CONTROL PANEL
14. WARNING HORN RELEASE
15. IGNITION CONTROL BOX
16. BOMB AND DROP TANK EMERGENCY RELEASE

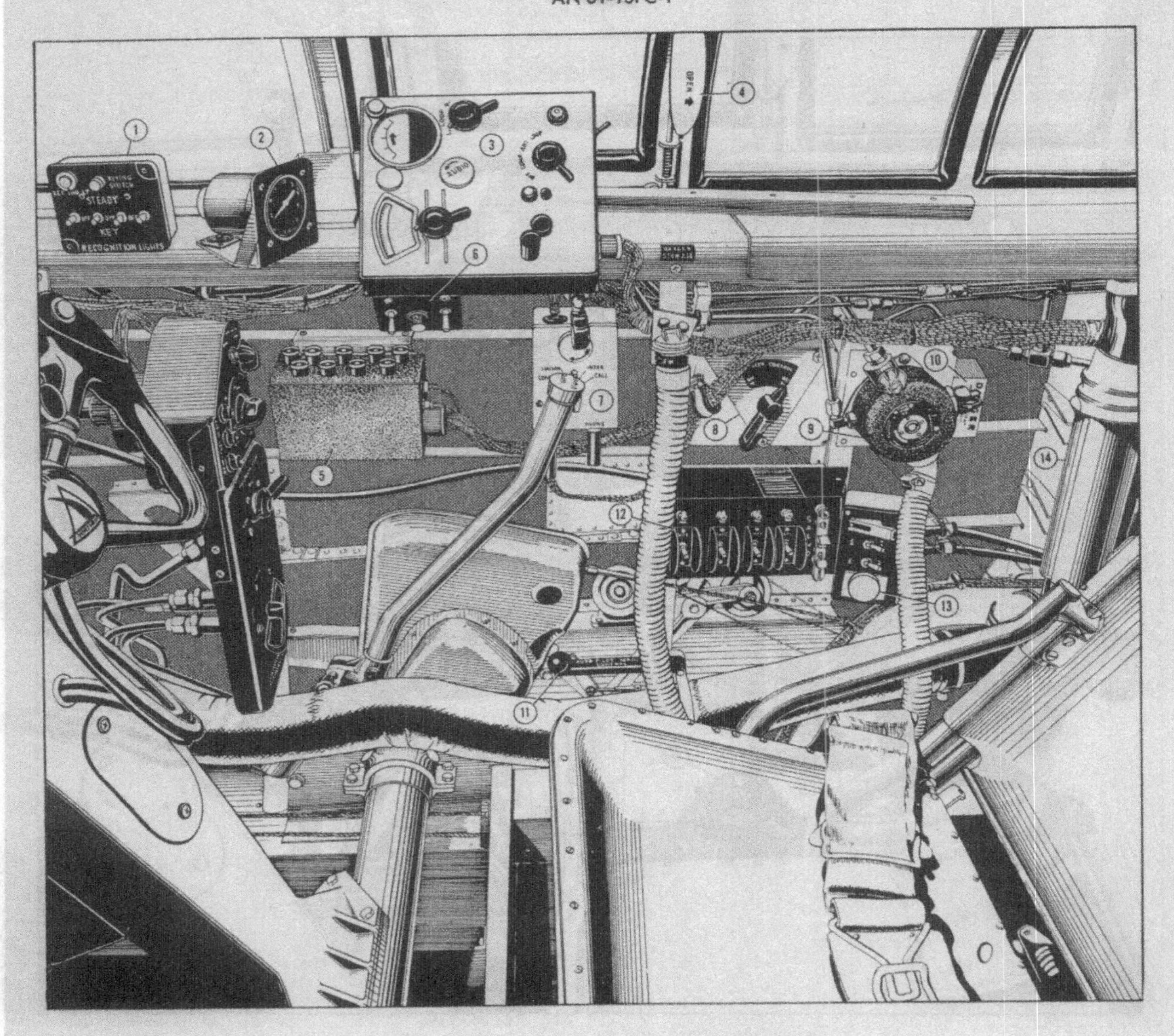

**Figure 8 — Pilot's Compartment — Right**

1. RECOGNITION LIGHT CONTROL BOX
2. RADIO COMPASS
3. RADIO COMPASS CONTROL BOX
4. WINDOW LATCH
5. AN/ARC-3 CONTROL BOX
6. AN/APS-13 CONTROL PANEL
7. PILOT'S BC-366 JACK BOX
8. DE-ICER CONTROL
9. PILOT'S OXYGEN REGULATOR
10. DESTRUCTOR
11. AUTOMATIC PILOT MASTER CONTROL
12. BOMB AND DROP TANK CONTROL BOX
13. SCR-695 CONTROL BOX
14. PILOT'S FLEXIBLE DEFROSTER TUBE

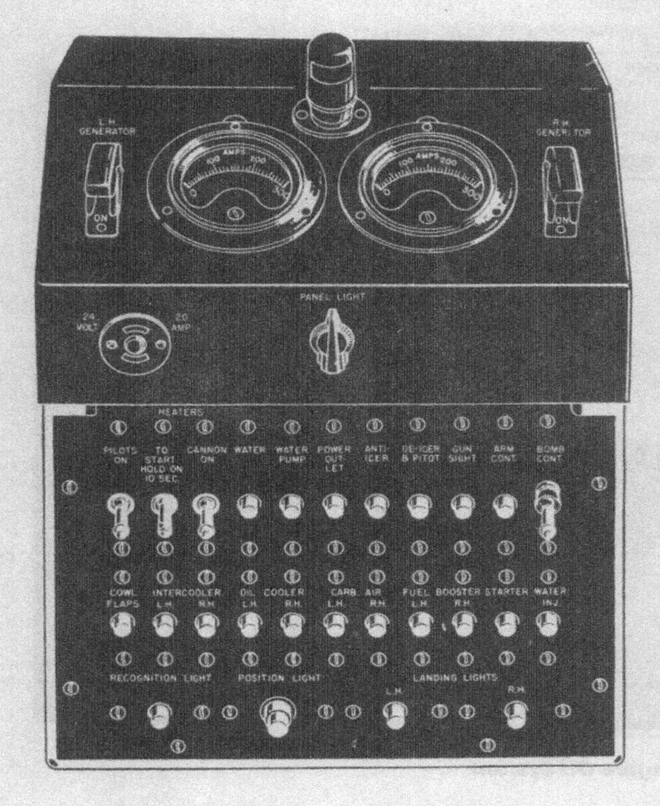


Figure 9 — Generator Control Panel

*d.* CROSS-FEED VALVE.—A cross-feed valve permits flow under pressure from one tank to another. For positioning of the cross feed valve control handle under various operating conditions, see figure 25.

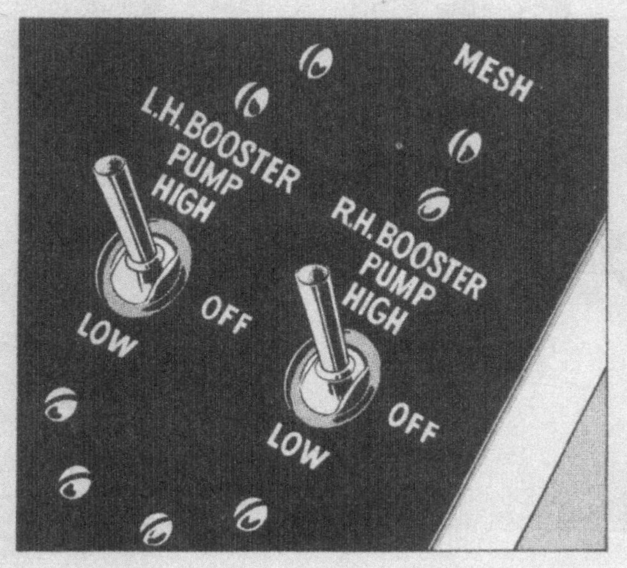


Figure 10 — Fuel Booster Pump Switches

Figure 11 — Propeller Control Box

## 5. INDUCTION SYSTEM CONTROLS.

(See figure 12.)

The induction system provides either ram air, heated air, or filtered air to the engine. Four-position switches, spring-loaded in the "HEAT" and "FILTER" positions and mounted on the auxiliary control panel, select the desired air condition. Indicator lights located above the switches, turn on when either the pre-heat or filter are operating. (See figure 16.) The air which becomes heated in passing through the turbosupercharger is cooled by ram air at the intercooler. The intercooler shutters may be set to operate automatically or may be controlled manually by a four-position switch,

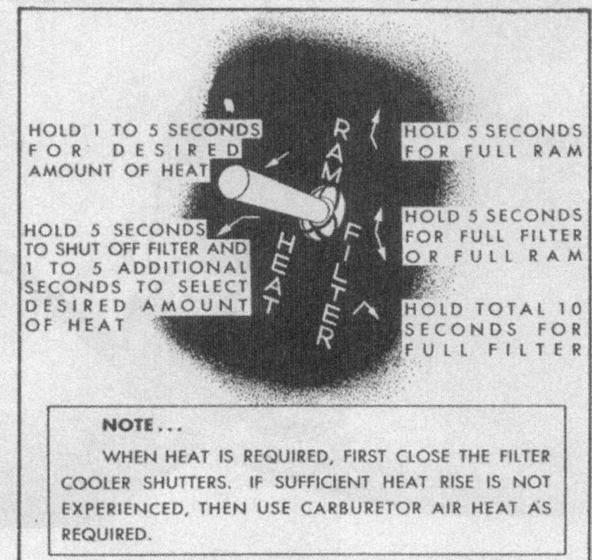


Figure 12 — Carburetor Air Control

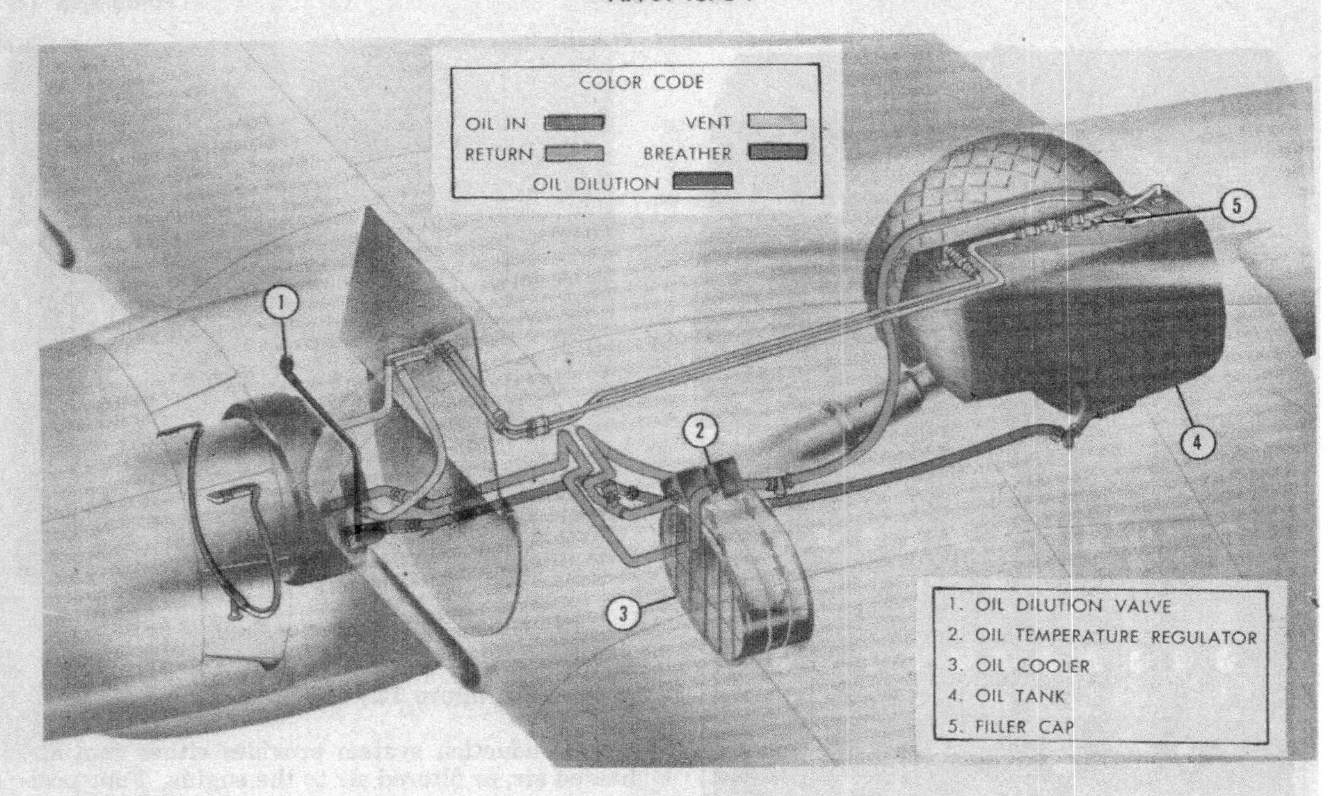


***Figure 13 — Engine Oil System***

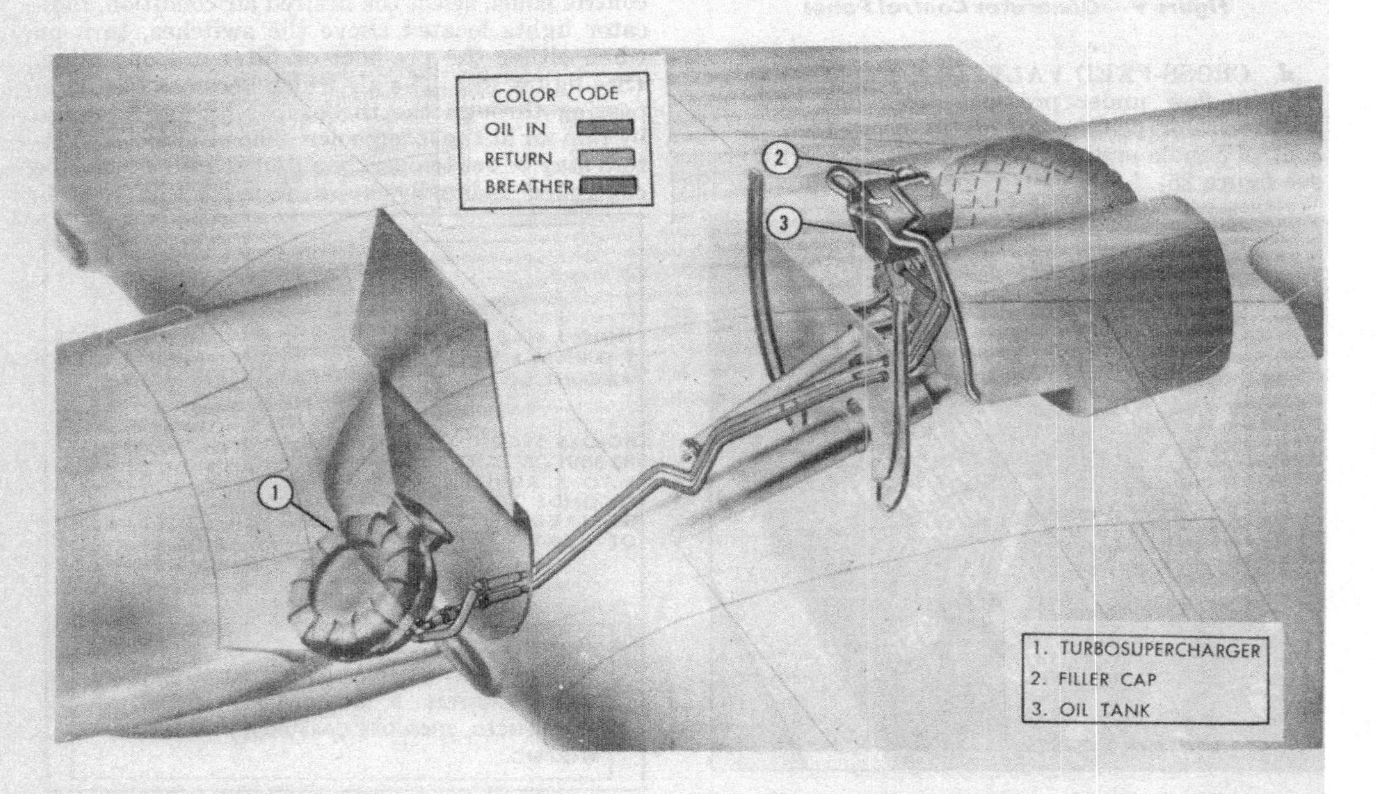


***Figure 14 — Turbosupercharger Oil System***

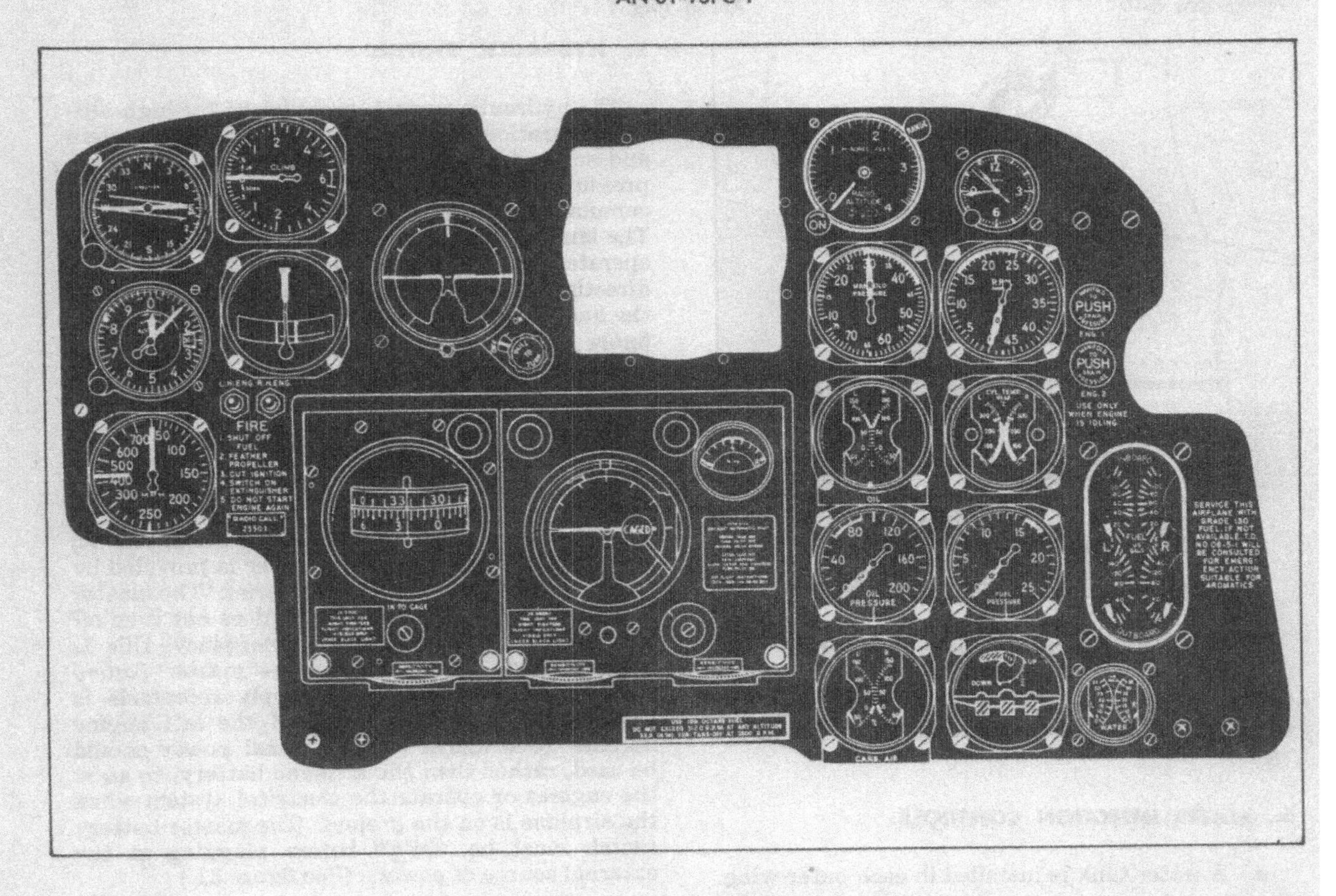


**Figure 15 — Pilot's Instrument Panel**

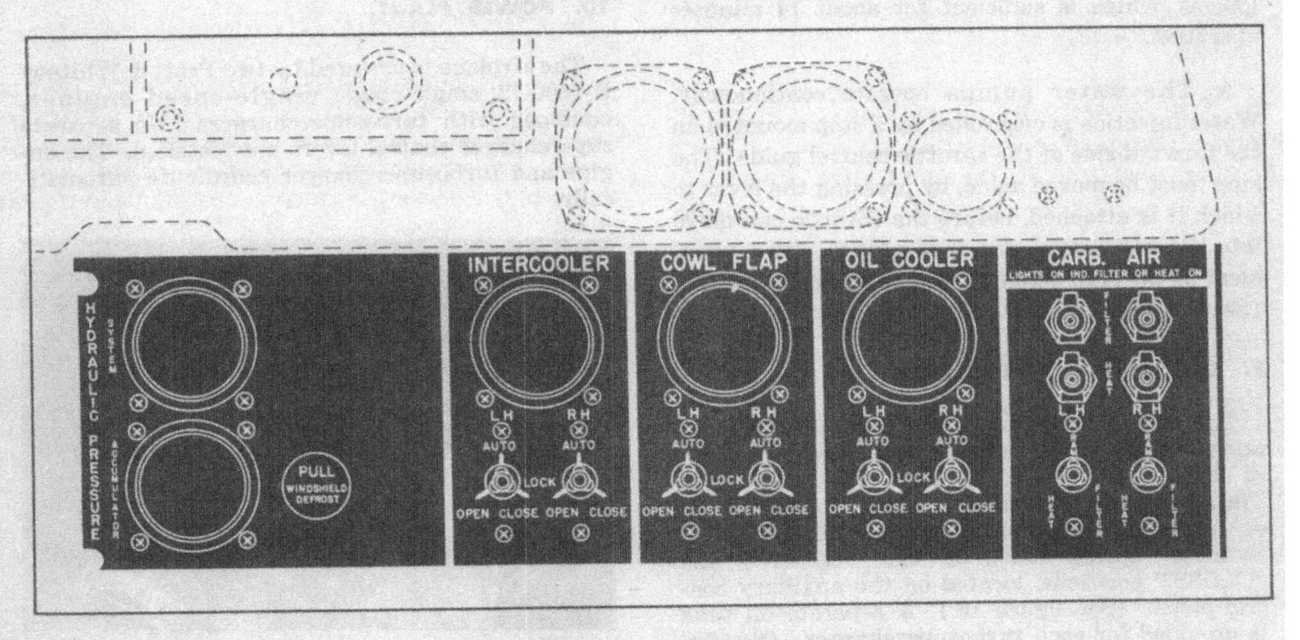


**Figure 16 — Auxiliary Instrument Panel**

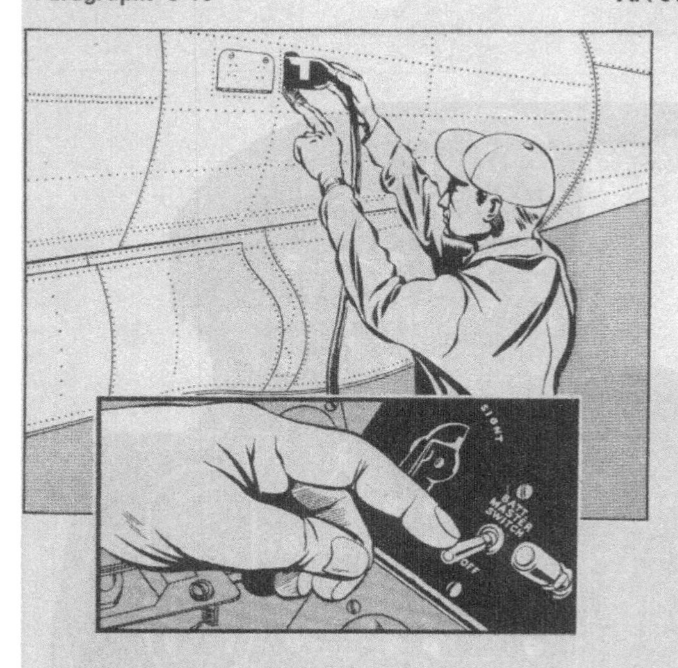


*Figure 17 — External Power Connection*

spring-loaded in the "OPEN" and "CLOSE" position, on the auxiliary instrument panel. (See figure 16.)

## 6. WATER INJECTION CONTROLS.

*a.* A water tank is installed in each outer wing. Each tank has a capacity of 32 US (26.6 Imperial) gallons, which is sufficient for about 14 minutes operation.

*b.* The water pumps operate continuously. Water injection is controlled by a stop mounted on the forward side of the throttle control guide. The stop must be moved aside, by pressing the lever to which it is attached, before the throttle can move into the war emergency power range. The movement of the stop actuates the water-power switch. (See 8, figure 7.)

## 7. OIL SYSTEM CONTROLS.

The oil tanks, pressurized for high-altitude operation and located in the wheel wells, have a capacity of 42 US (35 Imperial) gallons. (See figure 13.) The air flow through the oil cooler is automatically or manually controlled by two four-position switches, spring-loaded in the "OPEN" and "CLOSE" positions, located on the auxiliary control panel. (See figure 16.) A separate oil tank is provided for each turbosupercharger. (See figure 14.)

## 8. HYDRAULIC SYSTEM.

The hydraulic system, pressurized for high-altitude operation, consists of a main pressure system and an emergency hand-pump system. The main pressure system provides the pressure for the accumulator system and the automatic pilot system. The landing gear brakes and turret ejection doors operate from the accumulator, and the other units directly from the main pressure system. By use of the hand pump and hand-pump selector valve (see figure 37), fluid pressure can be directed to any hydraulically-operated unit except the automatic pilot.

## 9. ELECTRICAL SYSTEM.

A 24-volt, direct-current electrical system receives power from a generator, mounted on each engine. Supplementary power is provided by two 24-volt, 34-ampere-hour batteries. The master *ignition* switch (see 13, figure 6) does not turn off the electrical system. In an emergency, this is accomplished by turning off the master *battery* switch. An external power supply receptacle is located on the outboard side of the left engine nacelle. (See figure 17.) External power should be used, rather than the airplane battery, to start the engines or operate the electrical system when the airplane is on the ground. The master battery switch must be "OFF" before plugging in the external source of power. (See figure 31.)

## 10. POWER PLANT.

The airplane is powered by two Pratt & Whitney R-2800-73 single-stage, single-speed engines, equipped with turbosuperchargers. No separate supercharger control levers are installed. The engine and turbosupercharger coordinate automatically.

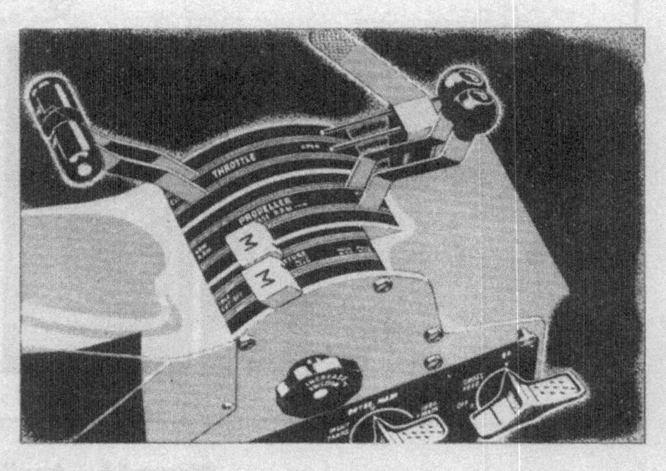


*Figure 18 — Engine Control Quadrant*

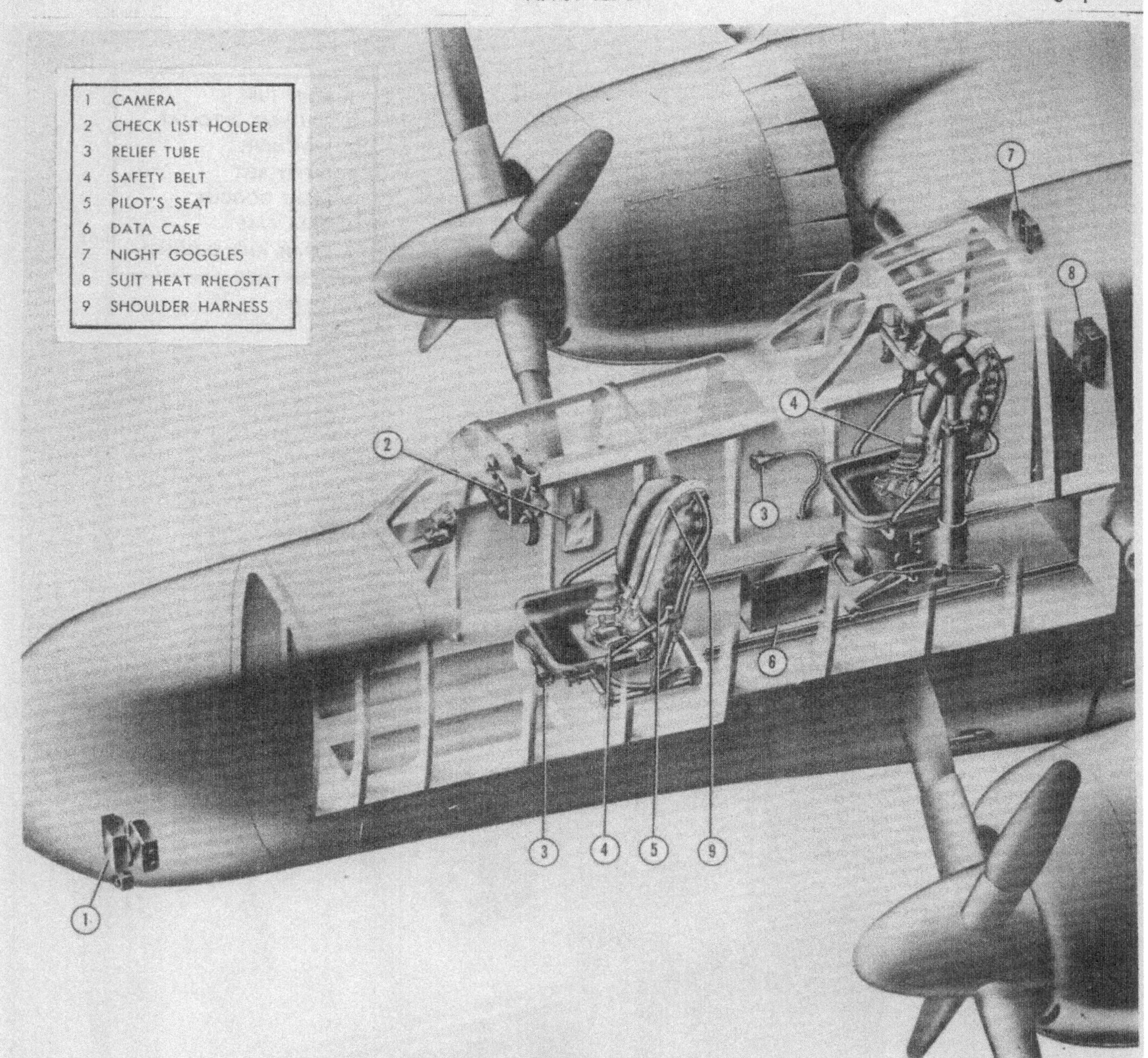


***Figure 19 — Crew Nacelle Equipment — Forward Compartment***

*a.* **POWER PLANT CONTROLS.**—Mounted on the engine-control quadrant are the throttle, mixture, and propeller-governor controls. (See figure 18.) The difference in the shape of the knobs on these controls makes it possible for the pilot to locate them in darkness, by touch alone. For immediate military power performance, throttle- and propeller-control levers can be pushed forward simultaneously with one hand.

(1) **THROTTLE.**—The throttle controls, mounted on the outboard side of the quadrant, are the longest pair of levers and have oval knobs. The throttle levers, in addition to providing throttle control, control the supercharger regulator.

(2) **PROPELLER.** — The propeller-governor controls, mounted in the center of the quadrant, are the medium-length pair of levers and have round knobs.

(3) **MIXTURE.**—The mixture controls, mounted on the inboard side of the quadrant, are the shortest pair of levers and have square knobs. Three mixture positions are provided: "IDLE CUT-OFF," "AUTO. LEAN.," and "AUTO. RICH."

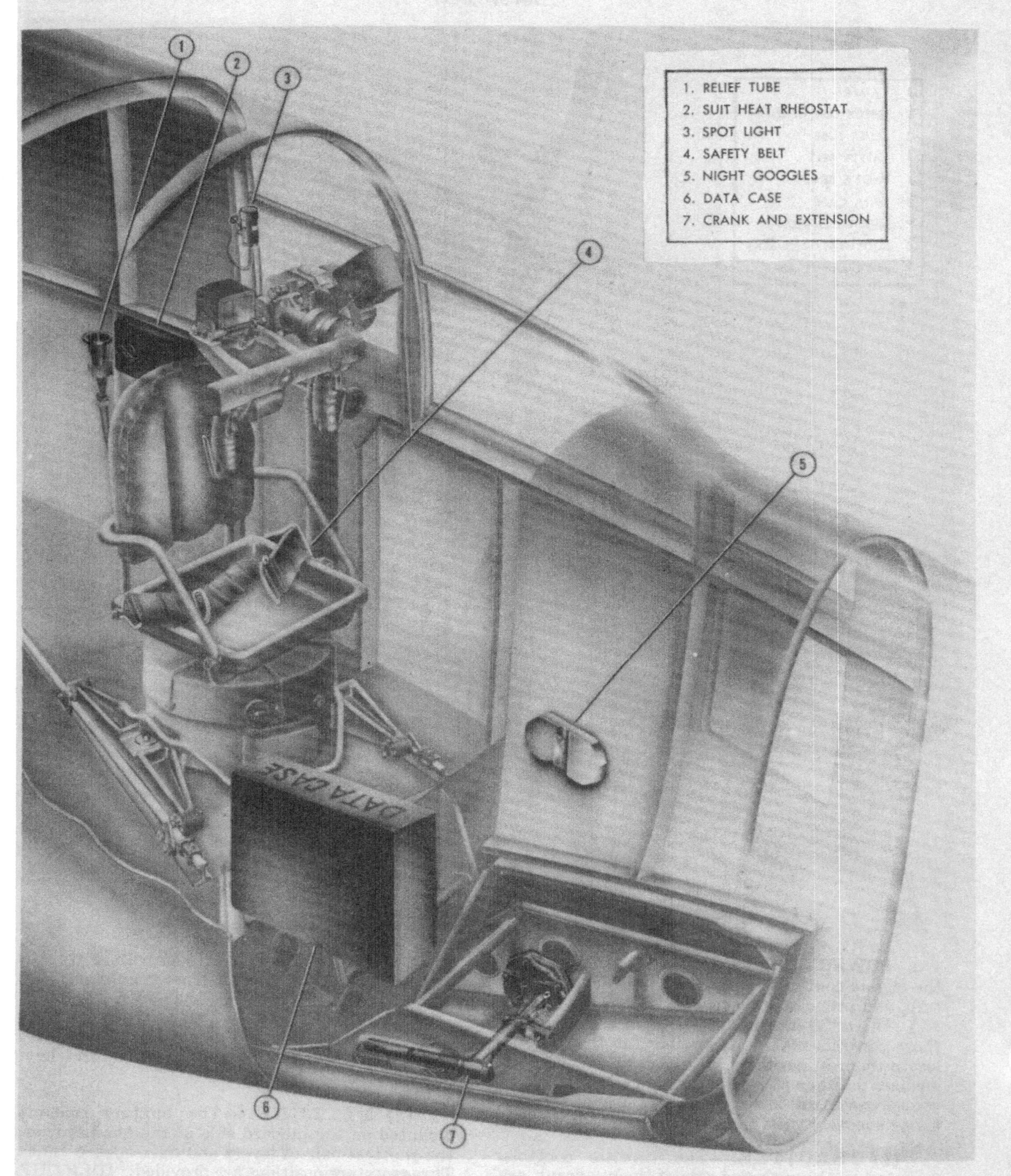


***Figure 20 — Crew Nacelle Equipment — Aft Compartment***

# SECTION II
# NORMAL OPERATING INSTRUCTIONS

## 1. BEFORE ENTERING THE PILOT'S COMPARTMENT.

*a.* RESTRICTIONS.—These limitations may be supplemented or superseded by instructions included in Service publications.

(1) FLIGHT RESTRICTIONS.—The following maneuvers are prohibited:

*(a)* Outside loops

*(b)* Inverted flight

*(c)* Spins

*(d)* Snap rolls

*(e)* Vertical reversement at speeds in excess of 250 mph IAS.

*(f)* All acrobatics and steep turns are prohibited when the drop tanks are full. When the tanks are full, restrictions on maximum vertical accelerations are as follows:

| | |
|---|---|
| Two 165 US (137 Imperial) gallon tanks | 5.06 g |
| Two 310 US (258 Imperial) gallon tanks | 4.00 g |
| Four 165 US (137 Imperial) gallon tanks | 4.65 g |
| Four 310 US (258 Imperial) gallon tanks | 4.00 g |

**Note**

With drop tanks empty, no maneuvering restrictions apply other than those of the basic P-61C airplane:

| | |
|---|---|
| Airplane clean | 5.67 g |
| Fighter brakes open | 4.00 g |

(2) AIRSPEED LIMITATIONS.

*(a)* Maximum allowable indicated airspeed, 430 mph IAS.

*(b)* Do not lower flaps FULL DOWN above 175 mph IAS.

*(c)* Do not operate the automatic pilot under 140 mph IAS.

*(d)* Do not lower landing gear above 175 mph IAS.

*(e)* Maximum diving rpm 3120.

*(f)* Emergency landing gear operation not above 130 mph IAS.

*(g)* Maximum permissible diving speeds with tanks installed, full or empty, are the same as for the basic airplane.

| Altitude | IAS |
|---|---|
| 0 to 10,000 feet | 430 mph |
| 10,000 to 20,000 feet | 375 mph |
| 20,000 to 30,000 feet | 305 mph |
| 30,000 to 35,000 feet | 270 mph |
| 35,000 to 40,000 feet | 240 mph |

*(h)* Maximum permissible indicated airspeeds at which empty tanks may be dropped in flight are:

| | |
|---|---|
| 165 US (137 Imperial) gallon tank | 350 mph IAS |
| 310 US (258 Imperial) gallon tank with 36-inch fin | 200 mph IAS |

(3) DO NOT OPERATE THE LANDING GEAR OR FLAPS WHEN THE AUTOMATIC PILOT IS ENGAGED.

(4) Continuous operation at war emergency power (water injection) is limited to five minutes.

(5) Continuous operation at military power is limited to 15 minutes.

*b.* TAKE-OFF GROSS WEIGHT AND BALANCE.—Check the airplane weight and balance chart.

*c.* ADDITIONAL CHECKS.

(1) PITOT TUBE: Cover removed.

(2) WING AIR DUCT COVERS: Removed.

(3) NOSE WHEEL LOCK PIN: Red cap on tight.

(4) BUNGEE PRESSURES: Gages for the main gear are located in the wheel wells; for the nose gear on the gunner's compartment, left-hand rail. (See 10, figure 64.) Main gear, 850 psi; nose gear, 700 psi.

(5) GUN BAY DOORS: Securely locked. Test with the wrench inside the 20-mm ammunition box door. Two distinct clicks can be felt and heard when each end of the gun-bay doors is locked.

*Figure 21 — Forward Compartments Entrance*

**Note**

The following checks must be made from the inner wing. Access to the inner wing may be gained by a padded ladder placed against the leading edge, or by way of the radio operator's escape hatch. If the escape hatch is used, be certain that it is securely locked when it is closed.

(6) OIL: Check quantity by dip stick. There is no oil-level gage. Also check the turbo-tank oil.

(7) FUEL, OIL, AND WATER TANK CAPS: Be certain they are firmly in place.

(8) GUN TURRET FAIRING: Make sure that it is latched and that all access doors are closed. Take hold of the guns and try to move them. They should be firmly fixed in the forward position.

*Figure 22 — Aft Compartment Entrance*

*d.* HOW TO GAIN ENTRANCE.

(1) The pilot and gunner enter their compartment by means of a ladder attached to the front entrance door, located in the nose-wheel well. (See figure 21.) The door is unlatched by a handle and opens down. To close the door, first step on the rod that folds the ladder, then pull the door up and turn the latch to secure it.

(2) The radio operator enters his compartment by means of a ladder attached to the rear entrance door in the bottom of the aft portion of the crew nacelle. (See figure 22.) The door is opened by pressing the large red button on the bottom of the nacelle just forward of the tail cone. To close the door, pull it up until it latches.

**2. ON ENTERING THE PILOT'S COMPARTMENT.**

*a.* STANDARD CHECK FOR ALL FLIGHTS.

(1) ALL ESCAPE HATCHES: *Make certain that all hatches are securely locked.*

(2) **TURRET COMPARTMENT CIRCUIT BREAKERS:** Switch on. (See figures 23 and 24.)

(3) **FORM 1A:** Check.

(4) **MASTER BATTERY SWITCH:** "OFF" if an external source of power is used; "ON" otherwise. Leave the master battery switch "OFF" until the external source of power is disconnected.

(5) **GENERATOR CONTROL PANEL:** Switches and circuit breakers "ON," except heater. If any large electric motor is heard running, check the radar and gun turret switches. They must be "OFF."

(6) **INSTRUMENTS:** Check for loose or broken cover glasses and proper indication:

*(a)* The wheel and flap indicator should show the wheels down and the flaps up.

*(b)* The fuel gage should agree with the entry in Form 1A. All tanks should be full.

*(c)* The oxygen gages should show 400-450 psi.

*(d)* The emergency air-brake-pressure gage should register 425 to 450 psi. The gage is located on the left-hand rail opposite the gunner's position.

*(e)* The hydraulic "SYSTEM" gage should register 850-1000 psi, and the "ACCUMULATOR" gage a minimum of 400 psi. If necessary, use the hand pump to bring up the accumulator pressure.

*(f)* **ANTI-ICING FLUID GAGE:** 5.75 US (4.8 Imperial) gallons.

(7) **FLIGHT CONTROL LOCK:** "OFF."

(8) **PARKING BRAKES:** Set.

(9) **TRIM TABS:** Movement free and full. Set the controls at neutral.

(10) **CROSS FEED VALVE CONTROLS:** "OFF."

(11) **FUEL TANK SELECTOR VALVE CONTROLS:** Set to "OUTBD. MAIN" tanks.

(12) **PROPELLER CONTROLS:** Push both circuit breakers down. Check feathering switches, "NORMAL." Check selector switches, "CONSTANT SPEED."

(13) **THROTTLE:** Approximately one-quarter open.

(14) **MIXTURE CONTROLS:** "IDLE CUT-OFF."

(15) **PROPELLER CONTROL LEVERS:** Full "INC. RPM."

***Figure 23 — Left-Hand Crew Nacelle Junction Box***

(16) **ELECTRICAL SWITCH PANEL:** Battery switches "ON," oil dilution "OFF," fuel booster pumps "LOW," pitot heater "OFF," position, landing, and taxi lights "OFF," camera "OFF."

(17) **LANDING GEAR CONTROL:** Firmly latched in the "DOWN" position.

(18) **COWL FLAPS:** Open the cowl flaps and return the switches to the "LOCK" position.

(19) **AUTOMATIC PILOT:** Disengaged.

(20) **VHF RADIO SWITCHES:** "OFF."

*b.* **SPECIAL CHECK FOR NIGHT FLIGHTS.**

**Note**

Wear night adapter goggles while testing lights.

(1) **WINDSHIELDS:** Absolutely clean. While it is always important that windshields be clean it is especially so for night flights. Dirty windshields reflect stray light and can cut down visibility as much as 50 per cent.

(2) **COMPARTMENT LIGHTS:** Test. There are four incandescent compartment lights; one over the propeller control panel, one over the radio, one on the canopy above the pilot's right shoulder, and one on the generator control panel.

(3) **FLOURESCENT LIGHT RHEOSTATS:** Check and adjust.

(4) **EXTERNAL LIGHTS:** Test operate
- Position lights
- Recognition lights (10 seconds maximum)
- Landing lights (10 seconds maximum)
- Taxiing lights

***Figure 24 — Right-Hand Crew Nacelle Junction Box***

**3. FUEL SYSTEM MANAGEMENT.**
(See figure 25 for examples.)

*a.* **CONTROLS.**—The fuel supply is directed to the engines by the fuel selector valve controls at the left of the pilot. (See 13, figure 7.) The forward selector handle controls the cross-feed valve, the middle selector handle controls the right main and drop tanks, and the aft selector controls the left main and drop tanks.

*b.* **NORMAL OPERATION.**

(1) Use fuel from the right-hand tanks for the right-hand engine and fuel from the left-hand tanks for the left-hand engine.

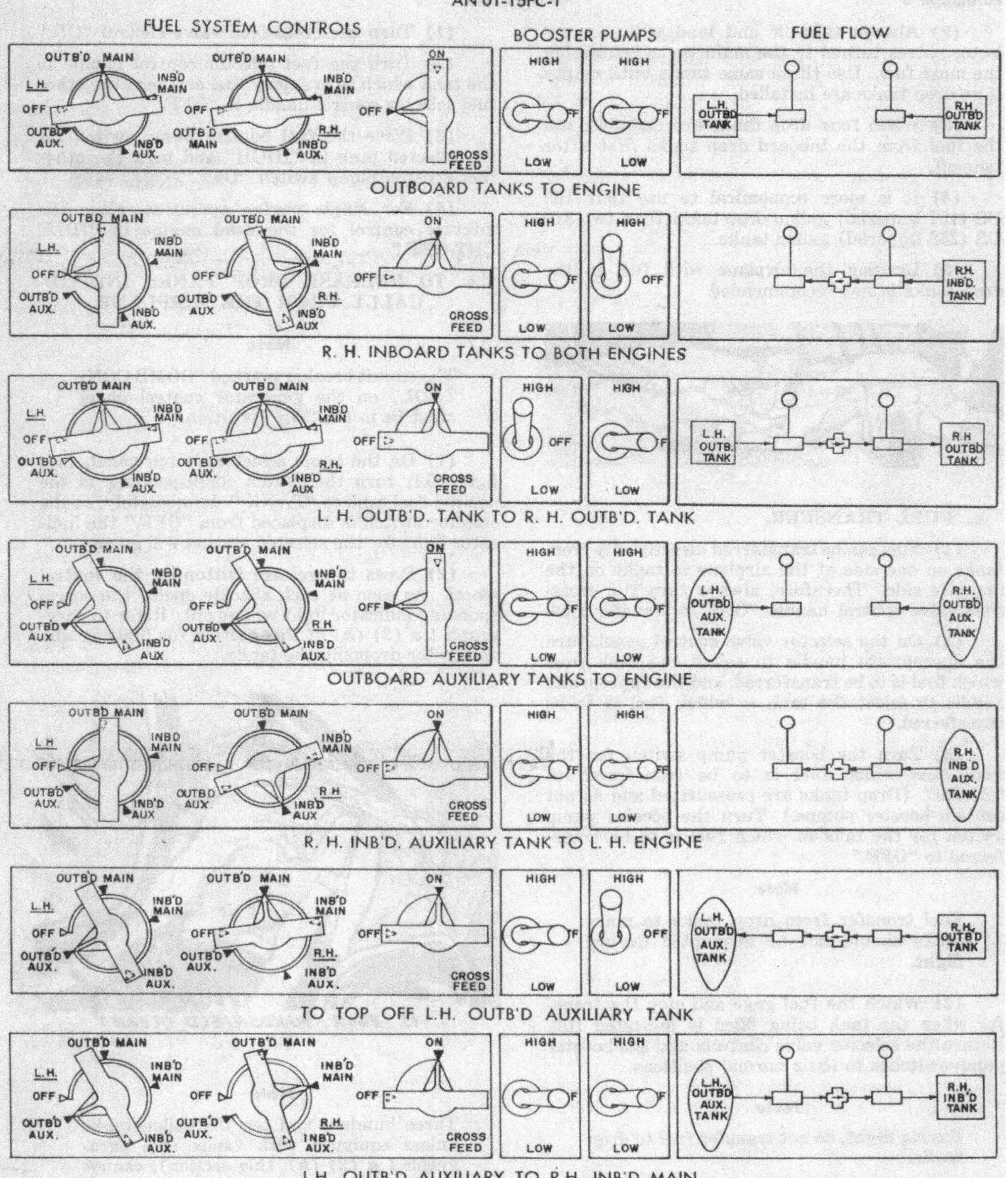


***Figure 25 — Fuel System Management Diagram***

(2) Always take-off and land with the selector valves turned to the main tanks containing the most fuel. Use these same tanks until empty if no drop tanks are installed.

(3) When four drop tanks are installed, use the fuel from the inboard drop tanks first after take-off.

(4) It is more economical to use four 165 US (137 Imperial) gallon drop tanks than two 310 US (258 Imperial) gallon tanks.

(5) Landing the airplane with fuel in the drop tanks is not recommended.

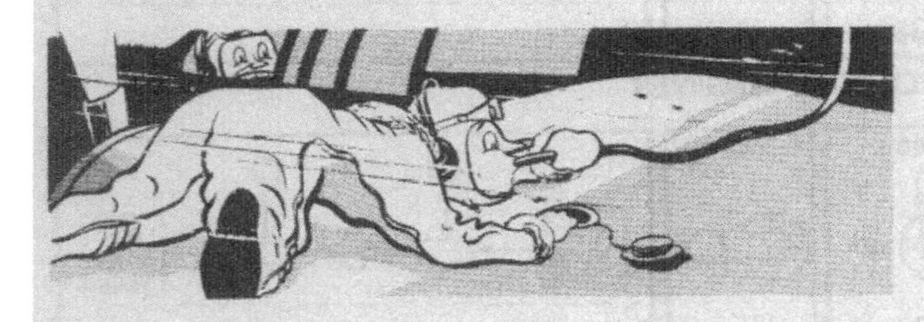

*c.* **FUEL TRANSFER.**

(1) Fuel can be transferred directly only from tanks on one side of the airplane to tanks on the opposite side. Therefore, always turn the cross-feed valve control handle "ON" to transfer fuel.

(2) On the selector valve control panel, turn the appropriate handle to select the tank *from* which fuel is to be transferred, and the appropriate handle to select the tank *to* which fuel is to be transferred.

(3) Turn the booster pump switch for the tank *from* which fuel is to be transferred to "HIGH." (Drop tanks are pressurized and do not contain booster pumps.) Turn the booster pump switch for the tank *to* which fuel is to be transferred to "OFF."

**Note**

Fuel transfer from drop tanks to main tanks should not be attempted during flight.

(4) Watch the fuel gage and stop the transfer when the tank being filled is indicated full. Return the selector valve controls and the booster pump switches to their normal positions.

**Note**

During flight, do not transfer fuel to drop tanks.

*d.* **EMERGENCY OPERATION.** — To operate two engines from one tank, or a single engine from a tank on the opposite side of the airplane:

(1) Turn the cross-feed valve control "ON."

(2) Turn the fuel selector control handle to the tank which is to supply fuel and turn the other fuel selector control handle to "OFF."

(3) Place the fuel booster pump switch for the selected tank in "HIGH" and turn the other fuel booster pump switch "OFF."

(4) For single engine operation, place the mixture control for the dead engine in "IDLE CUT-OFF."

*e.* **TO RELEASE DROP TANKS INDIVIDUALLY FROM THE AIRPLANE.**

**Note**

The circuit breaker marked "BOMB CONTROL" on the generator control panel must be in the "ON" position.

(1) On the bomb selector switch panel, (see figure 52) turn the switch corresponding to the desired fuel tank to "TANK." Immediately, as the selector switch is displaced from "OFF," the indicator light for the selected station will go on.

(2) Press the release button on the control wheel. As soon as each shackle opens, the corresponding indicator light will go out. Refer to paragraph **1** ***a*** **(2)** ***(h)*** of this section for limiting airspeeds for dropping the tanks.

*IS YOUR WINDSHIELD CLEAN?*

**Note**

Three hundred and ten US gallon tanks unless equipped with vanes (see paragraph **1** ***a*** **(2)** ***(h)***, this section), cannot be dropped, either singly or by "SALVO," without damage to the airplane unless sufficient fuel is left in the tank to keep it from "floating."

*f.* TO RELEASE TANKS BY SALVO.

**Note**

The circuit breaker marked "BOMB CONTROL" on the generator control panel must be in the "ON" position.

(1) On the bomb selector switch panel, turn the handle marked "SALVO" up. Immediately the indicator lights for all stations will go on.

(2) Press the release button on the control wheel. As soon as each shackle opens, the corresponding indicator light will go out.

**Note**

Indicator lights remaining on after the release button has been pressed show that the shackles have not opened. If a shackle opens but the tank fails to drop, stall the airplane. The slow speed and the buffeting may cause the tank to drop.

*g.* FUEL-VAPOR RETURN.—Vapor and fuel return from the carburetor to the main outboard tanks. The rate of return flow varies. If both carburetor vent floats are stuck, the return flow may be as great as 27 US gallons per hour.

**Note**

If a long flight is planned, after take-off and climb, turn the fuel selector controls successively to each pair of main and drop tanks for three or four minutes to make certain that all feed properly.

**4. STARTING ENGINES.**

*a.* INSTRUCTIONS IN CASE OF ENGINE FIRE.

(1) Allow the engine to continue running. The fire may be sucked through the engine and extinguished.

(2) Use the portable fire extinguishing equipment at once. A one-quart, carbon tetrachloride, hand fire extinguisher is located in each main gear wheel well. Access to the extinguishers is gained through doors, marked "FIRE EXTINGUISHER," in the outboard side of the wheel wells. These extinguishers should be removed and placed where they will be immediately available to members of the ground crew.

(3) If the fire persists, shut off the ignition and fuel supply.

(4) Use the fire extinguishing system, installed in each engine nacelle. The operating switch is on a panel below the pilot's compartment left rail, immediately aft of the propeller control panel. (See 2, figure 7.) In case of persistent fire:

*(a)* Close the cowl flaps in the burning engine nacelle.

*(b)* Place the fire extinguisher system switch in the on position for the affected engine.

*b.* DETAILED STARTING PROCEDURE.

**Note**

Either engine may be started first since the accessory equipment is the same for both.

(1) IGNITION SWITCHES: "OFF."

(2) WATER INJECTION SWITCHES: "OFF."

(3) PROPELLER: Pull the propeller through by hand sixteen blades in the direction of normal engine rotation if the engines have been idle for two hours or more. If the cylinders are loaded, remove the lower spark plugs and drain the fluid.

(4) INDUCTION SYSTEM AIR: "RAM" position. If dust is blowing, place the selector switch in the "FILTER" position.

(5) IGNITION SWITCHES: Master ignition switch "ON" and individual ignition switch "ON" for the engine to be started.

(6) FUEL BOOSTER PUMPS: "LOW."

**CAUTION**

While the booster pump is on, during the starting procedure, the mixture control should never be out of the "IDLE CUT-OFF" position for more than a second or two. Otherwise, the induction system becomes flooded, creating a fire hazard.

(7) STARTER SWITCH: "ENERGIZE."

**Note**

Energize the starter not more than 20 seconds when the airplane's batteries or a battery cart are employed. Energize not more than 10 seconds when an auxiliary generator is used. Do not exceed these limits because damage may result from over-speeding.

(8) **STARTER SWITCH:** "MESH."

**Note**

If the engine does not start the first time, allow the starter to cool before attempting another start. Be sure the starter jaws are disengaged before energizing the starter for another attempt. If the engine becomes loaded and fails to start promptly, advance the throttle to the full open position for three seconds with the mixture control in "IDLE CUT-OFF" and the starter turning the engine over.

(9) **PRIMER SWITCH:** On as necessary. If the engine is cold and has been exposed to outside temperatures below 15°C (60°F) priming will be necessary. The amount of priming varies from no priming for a warm engine to approximately ten seconds for a cold engine.

(10) **MIXTURE:** "AUTO-RICH," as soon as the engine fires.

(11) **THROTTLE:** Pull back quickly and smoothly and set for 700 rpm.

(12) **FUEL BOOSTER PUMPS:** "OFF."

(13) **FUEL PRESSURE:** 11 psi. Do not pump the throttles.

(14) **OIL PRESSURE:** If no oil pressure is indicated in 30 seconds, stop the engine and investigate the trouble. If oil pressure shows, start warming up the engine.

**Note**

If, for any reason, it is impossible to use the electric starters, a hand crank and extension are stowed in the radio operator's entrance door.

## 5. WARM-UP AND GROUND TEST.

*a.* **WARM-UP PROCEDURE.**

(1) Do not attempt to warm-up the engines more quickly by closing the cowl flaps or using carburetor air-heat.

(2) During ground operations at low temperatures when there are indications of carburetor ice, when the engines are rough, or when backfiring occurs, carburetor air-heat should be used to correct these conditions.

(3) Care should be exercised to avoid carburetor air temperatures above 50°C. Do not exceed 232°C cylinder head temperature.

(4) **THROTTLE:** 1000 rpm.

**Note**

Oil pressure may rise as high as 400 psi but will reduce as soon as the oil inlet temperature reaches 40°C.

(5) **COWL FLAPS:** Full open.

(6) **OIL COOLER DOORS.** Check operation. Oil temperature will rise more rapidly during warm-up if the oil cooler doors are closed.

(7) **HYDRAULIC AND ACCUMULATOR PRESSURE:** 850-1000 psi.

(8) **WING FLAPS:** Check operation.

(9) **DE-ICER PRESSURE:** 7-9 psi.

(10) **DE-ICER BOOTS:** Check operation (if installed).

(11) **AMMETERS:** Check.

(12) **RADIO EQUIPMENT:** Check operation.

*b.* **ACCESSORY CHECKS.**—Make the following checks *after* the oil inlet temperature has risen to 40° and the cylinder head temperature has risen to 130°C. Cowl flaps must be kept open for all ground operation. It is suggested that these checks be made on both engines simultaneously in order to save time.

(1) **MAGNETO CHECK.**

**Note**

The magneto check is more accurate if the engine is cleared before the check is made. Run the manifold pressure up to 40 inches Hg for five seconds, with the propeller control lever in full "INC. RPM" and the propeller selector switch in the "CONSTANT SPEED" position. Glance at the ammeter to be sure the generator is charging.

*(a)* CONTROL SETTINGS.

*1.* PROPELLER GOVERNOR: "Full RPM."

*2.* THROTTLE: 32 inches Hg.

**Note**

A throttle setting of 32 inches Hg should produce 2200 rpm. The magneto check must be made at an engine speed of not less than 2200 rpm.

3. OIL COOLER DOORS: "OPEN."

*(b)* PROCEDURE. — Turn each magneto switch from "BOTH" to "L" to "BOTH" to "R" to "BOTH." Allow a few seconds while at "BOTH" for the engine to pick up speed. Normal drop off is 50 to 75 rpm and must not exceed 100 rpm. If there is no drop in either magneto the switch is inoperative, and there is no way to check the magnetos.

**Note**

Freedom from engine vibration is a good indication of proper functioning, particularly of the ignition system.

**CAUTION**

Continued running at 25 inches Hg or above on one magneto may foul the plugs and will cause excessive fuel consumption.

(2) PROPELLER CHECK.—Supplementary to routine check made by the ground crew.

*(a)* AUTOMATIC OPERATION: Place the circuit breaker "ON," the selector switch in "AUTOMATIC," and the propeller control in take-off rpm. Open the throttle to obtain 2200 rpm, but do not exceed 32 inches Hg manifold pressure. Pull the propeller control back until the tachometer registers a drop of 200 rpm. Return the propeller control to the take-off position, noting that 2200 rpm is again attained.

*(b)* FEATHERING SWITCHES: Put the feathering switch in the "FEATHER" position. As soon as the tachometer shows a decrease in rpm, return the feathering switch to "NORMAL," noting that the original rpm is regained. The reduction in engine speed should not exceed 200 rpm. Close the throttle.

(3) FLIGHT INSTRUMENT CHECK.—Examine for proper setting and operation.

(4) AUTOMATIC PILOT GROUND TEST.—Check for proper operation.

(5) WATER INJECTION.

**Note**

This test must be made on both engines simultaneously. Operation of the stop lever on the throttle quadrant turns on the water to both engines. If either engine is running at less than 2200 rpm, turning on the water may kill the engine.

*(a)* Open both throttles to obtain approximately 2200 rpm with the propeller controls set at "HIGH RPM."

*(b)* After the engine speed has stabilized, press the lever at the front of the throttle quadrant to actuate the water injection power switch. The engine will hesitate momentarily and may become slightly rough when water is applied at this power.

*(c)* The water pressure gage should register a minimum of 20-22 psi (ground operation only).

(6) CHARGING 20-MM CANNON.

**WARNING**

The 20-mm cannon must not be charged until all pre-flight activities have been concluded, the pilot and crew members have entered their compartments, and the airplane is ready to take-off. This is a safety measure designed to protect personnel and property from accidental firing of the cannon while the airplane is on the ground.

*(a)* Remove the safety blocks (if installed) from the cannon breeches and charge the cannon.

(7) TAKE-OFF HORSEPOWER.—Hold the airplane with the brakes and check for full take-off horsepower (54.0 inches Hg at 2800 rpm).

6. SCRAMBLE TAKE-OFF.

**Note**

While taxiing dilute the oil if fluctuation in pressure is observed. Cowl flaps may be kept one-third open.

*a.* **MINIMUM VALUES FOR TAKE-OFF.**

(1) **OIL TEMPERATURE:** 40°C.

(2) **OIL PRESSURE:** 75 psi.

(3) **FUEL PRESSURE:** 17 psi.

(4) **MAGNETOS:** As soon as a normal magneto check (see paragraph 5 *b* (1), this section) can be made, it is permissible to take-off.

*b.* **POSITIONING OF CONTROLS.**

(1) **FUEL BOOSTER PUMP SWITCHES:** "HIGH."

(2) **WING FLAPS:** 20 degrees recommended if time permits.

(3) **THROTTLES:** Open gradually.

**7. TAXIING INSTRUCTIONS.**

*a.* Observe normal tricycle landing gear taxiing procedure.

*b.* **BATTERY AND GENERATOR SWITCHES:** "ON."

*c.* The maximum turning angle of the nose wheel is 50 degrees.

d. Taxiing in a crosswind, even a strong one, has little effect on the P-61. However, give the upwind engine enough extra power to offset the weathercock effect from the crosswind and to avoid riding the downwind brakes.

**8. TAKE-OFF.**

*a.* **NORMAL TAKE-OFF TECHNIQUE.**

(1) **CONTROL SETTINGS.**

*(a)* **TRIM TABS:** Neutral.

*(b)* **FUEL SELECTOR VALVE CONTROLS:** Outboard main tanks.

*(c)* **THROTTLE FRICTION LOCKS:** Sufficiently tight to keep controls from creeping.

*(d)* **WING FLAPS:** "UP" to 1/3 down, (0 degrees to 20 degrees) depending on length of runway and load.

**CAUTION**

Always set the wing flaps from the "UP" position to insure evenness.

*(e)* **COWL FLAPS:** Cracked open.

*(f)* **FUEL BOOSTER PUMPS:** "HIGH."

*(g)* **AUTOMATIC PILOT:** "OFF."

*(h)* **SURFACE CONTROLS:** Movement free.

*(i)* **GYRO INSTRUMENTS:** Uncaged and properly set.

*(j)* **COMPARTMENT HEATERS:** "OFF."

(k) **DE-ICER:** "OFF."

*(l)* **ANTI-ICER:** "OFF."

*(m)* **CROSS-FEED:** "OFF."

*(n)* **PROPELLER SELECTOR SWITCHES:** "CONSTANT SPEED."

*(o)* **PROPELLER GOVERNOR CONTROLS:** Full "INC. RPM" (2800 rpm).

*(p)* **MIXTURE:** "AUTO RICH."

*(q)* **INDUCTION CONTROLS:** "RAM," unless "FILTER" necessary.

(2) **INSTRUMENT READINGS.**

*(a)* **OIL TEMPERATURE:** 40-85°C.

*(b)* **OIL PRESSURE:** 75-95 psi.

*(c)* **CYLINDER HEAD TEMPERATURE:** 260°C maximum.

*(d)* **HYDRAULIC AND ACCUMULATOR PRESSURE:** 850-1000 psi.

*(e)* **RPM:** 2800.

*(f)* **MANIFOLD PRESSURE:** 54.0 inches Hg.

*(g)* **TORQUEMETER:** 118.6 psi (if installed).

(3) **BEST PROCEDURE.**

**Note**

Place the bomb "SALVO" switch in the salvo position in case it proves necessary to jettison the external load items. Replace the "SALVO" switch in the down position after a satisfactory take-off.

*(a)* Turn on to the runway. Open both throttles to about 40 inches Hg against the brakes for about 15 seconds to clear the engines and bring

the turbo up to speed. Release the brakes and smoothly apply full take-off throttle (53.5 inches Hg).

*b.* **MINIMUM RUN TAKE-OFF.**

(1) Start the take-off as close to the end of the runway as possible.

(2) Lower the landing flaps 1/3 (20 degrees).

(3) Run the engines to full take-off manifold pressure (54.0 inches Hg) against the brakes.

(4) Release the brakes and start the run, holding the nose wheel on the ground while picking up speed.

(5) Pull the nosewheel off the ground and take off as soon as flying speed is attained (110 mph at 30,000 lbs. gross weight.)

(6) Retract the landing gear and level off to attain critical single engine speed before climbing.

**9. ENGINE FAILURE DURING TAKE-OFF.**

The pilot must determine quickly whether or not sufficient flying speed has been attained for single engine flight. This will vary with the gross weight at time of engine failure.

*a.* **IF ONE ENGINE FAILS AND THERE IS INSUFFICIENT SPEED FOR SINGLE ENGINE OPERATION:**

(1) If drop tanks are installed drop them as quickly as possible whether the airplane has left the ground or not. If bombs are installed drop them "SAFE."

(2) If one engine fails before the airplane has left the ground, cut the power on both engines and stop straight ahead. If it is apparent that the airplane will run off the runway before it can be brought to a stop, use maximum brakes without skidding the tires until near the end of the runway, then retract the gear.

(3) If one engine fails in the air before the airplane has attained critical single engine speed, cut the power on both engines and land straight ahead. If insufficient runway is available for landing retract the landing gear at once.

(4) Never try to turn more than 15 or 20 degrees in attempting to land straight after one engine has failed during take-off.

*b.* **IF SUFFICIENT FLYING SPEED HAS BEEN ATTAINED FOR SINGLE ENGINE OPERATION.**

(1) **LANDING GEAR:** "UP."

(2) **DROP TANKS:** Release immediately.

(3) **DEAD ENGINE FEATHERING PROCEDURE.**

*(a)* Close the throttle.

*(b)* Set the feathering switch to "FEATHER."

*(c)* Move the mixture control to "IDLE CUT-OFF."

*(d)* Shut off the fuel supply to the idle engine.

*(e)* Turn the ignition switch "OFF," after the propeller stops rotating.

*(f)* If, for any reason, the feathering circuit fails to work, return the feathering switch to the normal position and hold the propeller selector switch in "DEC. RPM." until the blade is feathered.

*(g)* Manually "CLOSE" the cowl flaps.

**Note**

In the event the propeller still fails to feather and remains in fixed pitch, windmill the propeller at the lowest possible rpm. The propeller will windmill at a speed proportional to the airspeed, making it desirable to fly the airplane at an airspeed not more than 20 to 30 mph above the stalling speed. Place the controls for the inoperative engine in the positions for feathering as given above.

(4) **FLAPS:** "UP" when an altitude of 500 feet has been attained.

(5) **RUDDER TRIM TABS:** Adjust for directional control.

(6) **TURNS:** Do not turn until trim is established.

**Note**

The airplane has a high acceleration just after leaving the ground. When in the air, it has exceptionally fine single engine performance.

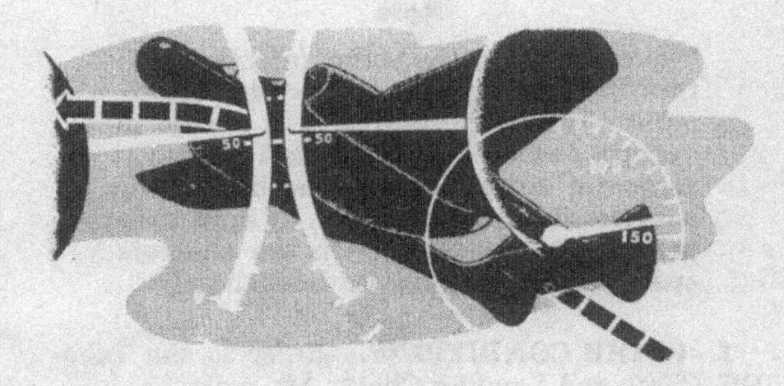

## 10. CLIMB.

The speed for best rate of climb is approximately 150 mph IAS. During climbs, cruising, and maneuvers the engine speed should be regulated by the propeller control levers; the propeller control switches should be in the "AUTOMATIC" position. Never begin the climb until critical single engine speed has been attained.

*a.* LANDING GEAR: "UP."

*b.* WING FLAPS: "UP" as soon as possible after attaining critical single engine speed. The flaps may not lock in the up position if they are raised at over 175 mph IAS.

*c.* COWL FLAPS: As desired.

*d.* FUEL BOOSTER PUMPS: "OFF" unless the climb is to continue above 15,000 feet, or unless the engine driven pumps will not maintain 17-19 psi. In the latter case, land and have the situation checked.

*e.* CLIMBING.

(1) Normally, maintain a speed of 165 mph IAS, up to about 20,000 feet. Long climbs at lower IAS will overheat the engines, except in cold weather.

(2) Above 20,000 feet, reduce the IAS somewhat, varying it according to the oil and cylinder head temperatures registered. Never allow the oil temperature to rise above 110°C, the cylinder head temperature to exceed 260°C, or the carburetor air temperature to exceed 50°C.

(3) In a short, rapid climb, maximum rate can be obtained at 150 mph IAS. This is for short climbs only. Save it for emergency use. Lower IAS and full power will cause oil and cylinder head temperatures to rise swiftly.

(4) In lower power climbs use "AUTO RICH," if necessary, to keep cylinder head temperatures low.

**Note**

When operating the water injection, full throttle may be used up to 54.4°C carburetor air temperature. It is unsafe to operate above 54.4°C. If limit temperatures are obtained, open the intercooler shutters full open manually. If limit temperature is exceeded, with the shutters open, retard the throttle immediately.

*f.* CLIMB CONDITIONS.—Refer to the Take-Off, Climb and Landing Chart, Appendix I.

## 11. GENERAL FLYING CHARACTERISTICS.

*a.* MAXIMUM PERFORMANCE.

(1) The maximum performance of the airplane can be obtained only by operating at the limiting conditions in every factor. Weight should be kept at a minimum for the particular mission and the pilot must be familiar with the handling characteristics at maximum performance.

(2) Pilots should not hesitate, in order to obtain maximum performance, to operate at 260°C cylinder head temperature. Cowl flaps should be kept fully closed as long as the above, maximum temperatures are not exceeded. It is obvious that the high speed capabilities of the airplane will not be realized if an over-conservative attitude is held relative to temperature ranges, and if cooling flaps and shutters are allowed to remain open to create performance killing drag.

(3) For continuous operation maximum cylinder head temperature is 232°C and maximum oil temperature is 85°C.

(4) In level flight it is permissible to operate with the mixture control in "AUTO. LEAN" up to normal rated power (2600 rpm and 42.0 inches Hg) provided cylinder head temperatures do not exceed 232°C and carburetor air temperatures do not exceed 50°C. The use of powers higher than normal rated are not permitted in "AUTO. LEAN" under any circumstances. The use of "AUTO. LEAN" at higher powers than cruising should be confined to conditions where range in conjunction with high powers is a factor.

*b.* OPTIMUM CRUISING.

(1) MAXIMUM RANGE. — Maximum range will be obtained at low altitudes. The throttles should be adjusted to 33 inches Hg with "AUTO. LEAN" mixture. Adjust the rpm to maintain an IAS of about 170 mph. The lowest practical rpm setting is 1400. To obtain maximum range all the fuel in the drop tanks must be used and they must be dropped in pairs.

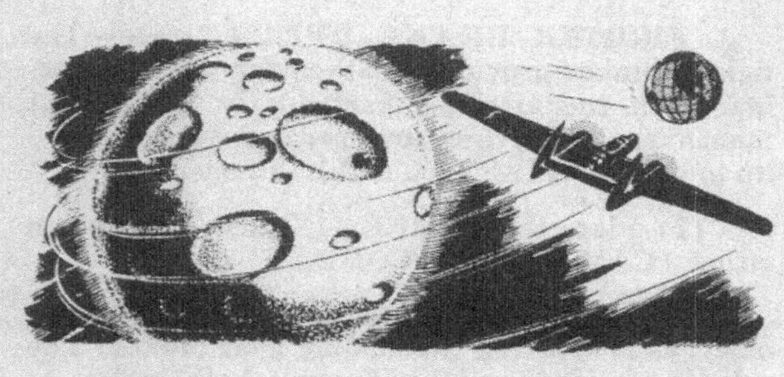

(2) **MAXIMUM ENDURANCE.**—To remain in the air as long as possible fly at an IAS of about 160 with 31.5 inch Hg manifold pressure. Maximum endurance will be at sea level.

*c.* **CHANGING POWER CONDITIONS DURING FLIGHT.**

(1) **TO INCREASE POWER.**

*(a)* Increase the engine rpm.

*(b)* Adjust the throttle to the desired manifold pressure.

**Note**

No separate supercharger adjustment is possible; operation is automatic. A few seconds lag in manifold pressure build-up, as the supercharger rpm increases to the new output level, can be noted.

**Note**

Do *not* increase the manifold pressure prior to increasing the rpm.

(2) **TO REDUCE POWER.**

*(a)* Adjust the throttle to the desired manifold pressure.

*(b)* Decrease engine rpm.

**Note**

Do *not* decrease the engine rpm before reducing the manifold pressure.

*d.* **PROPELLER OPERATION.**

(1) **MANUAL CONTROL.** — Manual control is to be used in flight only in the event of failure of the automatic control, or if it is necessary to conserve electric power due to generator failure. When the selector switch is in the "FIXED PITCH" position the engine rpm will vary with airspeed and power.

(2) **FEATHERING AND UNFEATHERING IN FLIGHT.**

*(a)* **FEATHERING.** See section II, paragraph 9 *b* (3).

*(b)* **UNFEATHERING.**

*1.* See that the throttle is closed and the mixture control is in "IDLE CUT-OFF."

*2.* Turn the ignition switch "ON."

*3.* Place the propeller control in the "DEC. RPM" position.

*4.* Turn on the fuel supply.

*5.* Place the feathering switch in the normal position and hold the selector switch in "INC. RPM" until the engine speed reaches 800 rpm, then place it in the fixed pitch position.

*6.* Move the mixture control to "AUTO. RICH," and place the throttle control in the normal starting position.

*7.* As soon as the proper engine operating temperatures are reached, place the selector switch in "CONSTANT SPEED." Adjust the mixture, throttle, and propeller control levers to the desired power and engine rpm.

*e.* **AUTOMATIC PILOT OPERATION.**

(1) **SENSITIVITY CONTROL SETTING.**—Best setting of these dials will vary, but the following settings are recommended:

| | |
|---|---|
| Rudder | between "AVERAGE" and "SMOOTH" |
| Elevator | "AVERAGE" |
| Aileron | "AVERAGE" |

(2) **ENGAGING THE AUTOMATIC PILOT IN FLIGHT.**—After aligning the indices and uncaging the gyros, turn on the "AUTOMATIC PILOT OIL PRESSURE VALVE" lever (see 14, figure 6) and then engage the "AUTOMATIC PILOT MASTER CONTROL" (see 11, figure 8). (Reverse the procedure when disengaging the automatic pilot.) By holding the controls, the pilot can feel the automatic pilot taking over.

*f.* **SUPERCHARGER OPERATION.** — Completely automatic. However, a brief time lag will be noticed on increasing supercharger rpm in certain power conditions.

(1) **TURBOSUPERCHARGER CHARACTERISTICS.** *(See figure 26.)* — Two phenomena occur with turbine operation, which may cause those unfamiliar with them some concern:

*(a)* One is turbine collapse. This may occur at intermediate altitudes when power is decreased by reducing engine speed while holding a relatively high manifold pressure. If the wastegate is nearly closed, a point may be reached where there is insufficient quantity of exhaust gas to maintain the turbine speed required to hold the manifold pressure. This results in reduced turbine speed with a reduction of manifold pressure, which further re-

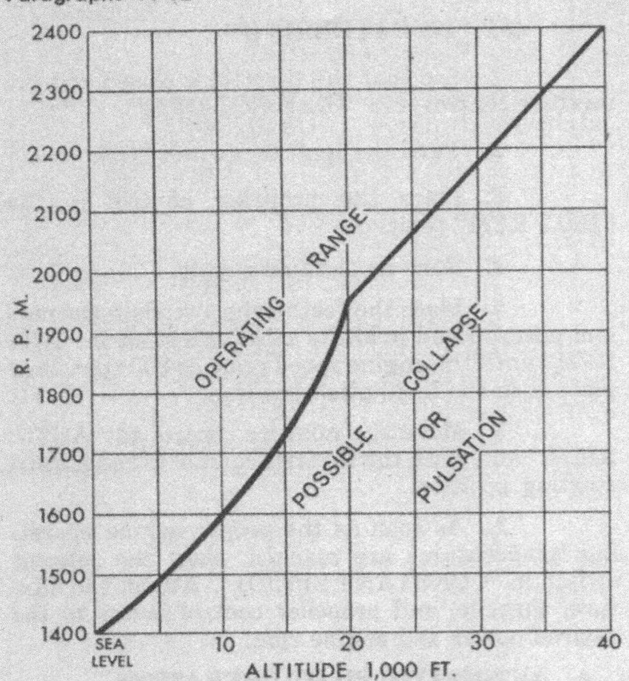


***Figure 26 — Turbo Operating Range***

duces engine power so that a slow "collapse" of the power system occurs. This phenomenon may be recognized by a gradual lowering of the manifold pressure with no change in throttle control position. In order to prevent complete collapse, and stopping of the engine, the engine speed (rpm) should be increased, and the power reduced, if desired, by reducing the manifold pressure.

*(b)* Pulsation is the second phenomenon. It is more likely to occur than "collapse," and is a characteristic of highly turbosupercharged engine installations, operating at high altitudes. It is caused by a stalling of the compressor-impeller and diffuser blades. It occurs when power is reduced by reducing engine speed while holding a relatively high manifold pressure, thereby creating a condition where the engine cannot use all the air the supercharger attempts to pump. Pulsation can be identified by violent fluctuations of the manifold pressure and intermittent misfiring. It is not harmful to the engine, but if the manifold pressure reaches too low a value, the engine may quit completely. The remedy is the same as that for cases of turbine collapse—increase the rpm, and reduce power, if desired, by use of the throttle lever.

*(c)* Since it is generally more economical to operate at reduced powers by operating at the lowest possible rpm and highest allowable manifold pressure, it is important, where economy is desired, to operate as close to the pulsation limits (where they are limiting factors) as possible. In order to aid the pilot, a curve of the approximate rpm, where either pulsation or collapse may occur, vs. altitude is shown in figure 26.

*g.* **FIGHTER BRAKE OPERATION.**—The fighter brakes provide a method for rapidly slowing down the airplane, particularly for combat maneuvers. They may be operated at any speed up to placarded IAS (430 mph). To operate:

**(1)** Place the CAMERA-COMBAT switch in either "CAMERA" or "COMBAT" position.

**(2)** Operate both brake pedals, simultaneously, as if applying the landing gear brakes. The release of either pedal will cause both brakes to close.

**(3)** A slight nose-up tendency is evident when the brakes are opened. This is readily controlled with a little experience.

**(4)** Deceleration will be rapid and immediate.

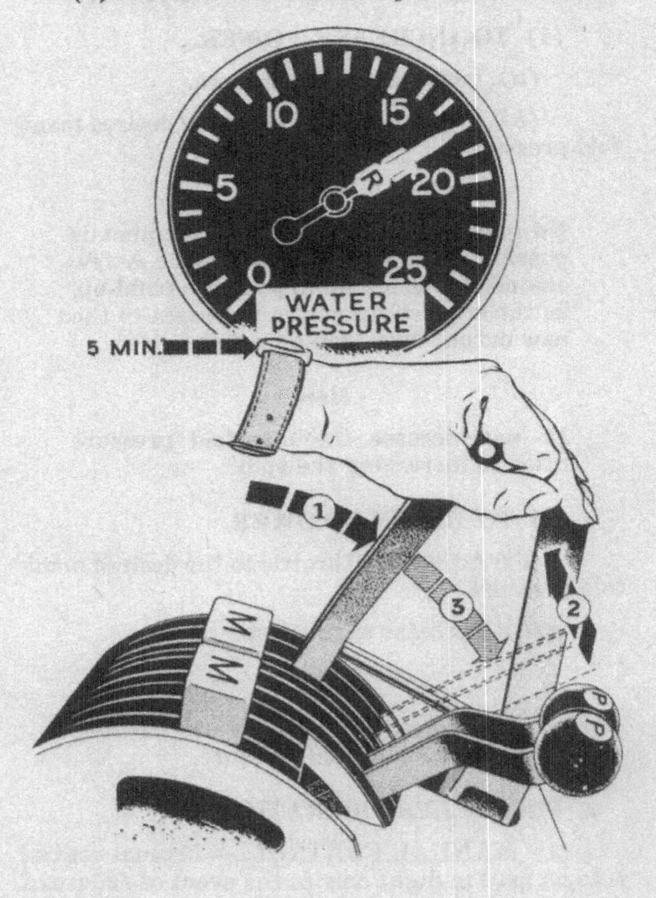


***Figure 27 — War Emergency Operation***

## 12. WAR EMERGENCY OPERATION.

*a.* Set the propeller control to full "INC. RPM."

*b.* Advance the throttles to the stop.

*c.* Move the stop out of the way with the thumb and advance the throttles to the full "OPEN" position.

**Note**

When the water injection is turned on a lapse of a few seconds will occur before it takes hold. The effect is gradual rather than immediate.

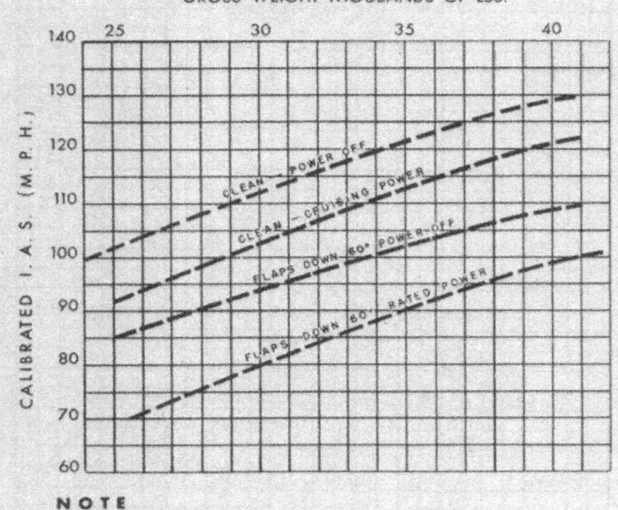


*Figure 28 — Stalling Speed Chart*

## 13. STALLS.

*a.* **APPROXIMATE STALLING SPEEDS.** (See figure 28.)

*b.* **STALL WARNING.**—Ample warning of an impending stall occurs in the form of strong tail buffeting.

*c.* **EFFECTIVENESS OF CONTROLS.**

(1) Up to and in the stall, elevator, rudder, and aileron controls remain effective. Deflection of the controls is unnecessary to prevent roll in the stall once the controls have been set to produce a stall.

(2) No aileron deflection is necessary to hold the airplane level laterally at cruising power in the stalled condition. At military power, five to ten per cent aileron deflection is required. With gear and flaps down, 75 to 90 per cent aileron deflection is necessary.

(3) In stalls occurring with one engine inoperative, rudder forces to compensate for yaw are high, requiring 90 to 100 per cent full rudder deflection.

(4) The airplane can be stalled at high speed in a turn. It has no tendency to whip in either direction, but does a series of stalls and recoveries while continuing to turn.

(5) The airplane recovers straight ahead from a stall. There is no tendency to fall off to one side or the other, whether the flaps and landing gear are down or up.

**Note**

Because the airplane has such unusually excellent stall characteristics, it is recommended that pilots investigate normal and single-engine stalls. Use caution to maintain sufficient altitude because the rate of descent in stalls is exceedingly high.

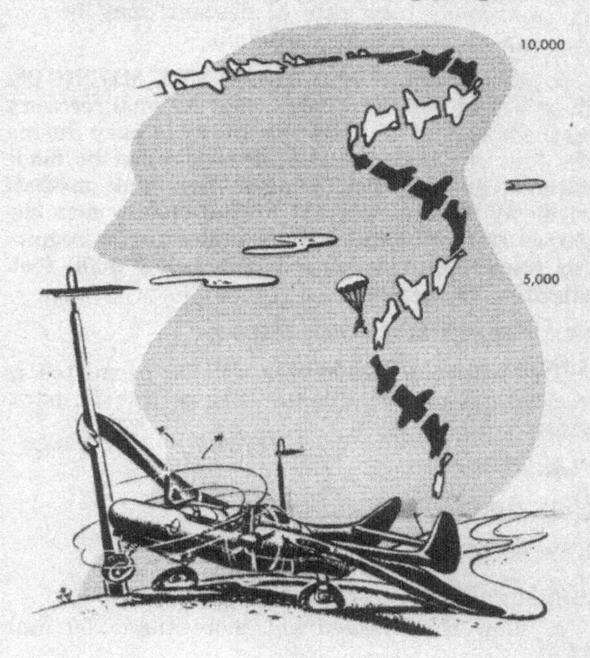


## 14. SPINS.

*a.* **SPINNING CHARACTERISTICS AND LIMITATIONS.**

(1) There is no tendency for the airplane to spin inadvertently in either the cruising or landing attitude. In fact, it is extremely difficult to spin the P-61 airplane. It takes approximately ¾ of a turn, with full elevator and full rudder applied, to start spinning. Recovery may be made immediately during the first ¾ of a turn by relaxing pressure on either the rudder or elevator control. After that, the spin tightens and acceleration increases. The airplane will lose approximately 2000 feet per turn.

(2) With the landing gear and flaps up, the airplane falls off slowly and deliberately. At the end of an 180-degree turn, the nose of the airplane will be nearly straight down. Rate of descent is

high, although forward velocity does not increase appreciably during a half turn.

(3) With the landing gear and flaps down, spinning characteristics are similar to those of the airplane in the clean condition. In addition, continuous and fairly severe buffeting occurs and the spin is oscillatory in nature. Recovery is considerably slower than when the airplane is in the clean condition.

(4) Intentional spins are prohibited.

**CAUTION**

Very high control forces, coupled with an extremely high rate of descent, may be expected in a developed spin.

*b.* **RECOMMENDED RECOVERY METHODS.** In case of an accidental spin, conventional recovery methods should be used. Recovery is easy during the first one and one-half turns and can be made without great difficulty after two and one-half turns by abrupt reversal of the rudder and elevator. If there is no definite indication of regaining control in a developed spin below 5000 feet, abandon the airplane.

## 15. PERMISSIBLE ACROBATICS.

The following maneuvers will be permitted in the normal load condition. (See section II, paragraph 1 *a*.)

*a.* Half roll

*b.* Normal loop.

*c.* Immelman

*d.* Chandelle

*e.* Slow rolls (speed not more than 250 mph IAS)

*f.* Vertical reversement (speed not more than 250 mph IAS)

*g.* Barrel roll

*h.* Intentional stall

## 16. DIVING. (See figure 29.)

*a.* **DIVING CHARACTERISTICS AND LIMITATIONS.**

(1) Unless normal rated power is being used for tactical purposes, dives should be started with the rpm and manifold pressure set at some cruising value. This will cause the governor to act with less time lag, as the airspeed increases, than if the take-off setting of the propeller governor were used.

(2) During dives the mixture controls should be in the "AUTO. RICH" position.

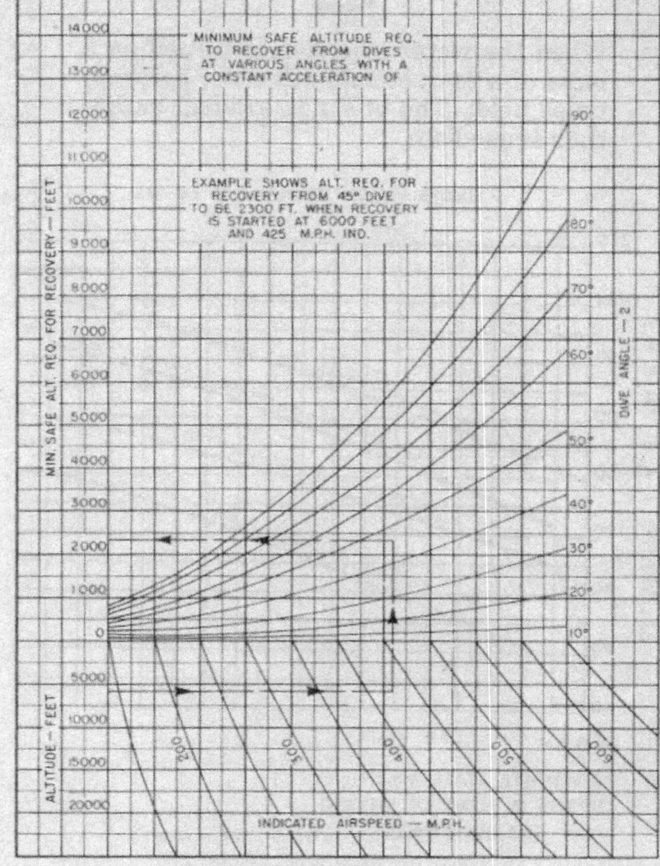


*Figure 29 — Dive Chart*

(3) Once the airplane has been trimmed for an airspeed of 275 mph no further trimming should be necessary.

*b.* **HIGH-SPEED DIVES AND STALLS.** — In high-speed dives and accelerated stalls compressibility affects in the form of buffeting and a tendency toward nose heaviness occurs at a speed approximately 70 per cent of the speed of sound. The approximate variation of this speed with altitude is shown by the diving-speed limitations in paragraph 1 *a* (7), this section. A pullout of at least 2.5 gs can still be made after this speed has been reached. Buffeting occurs at about 2 gs during the pullouts at speeds above approximately 60 per cent of the speed of sound, but the airplane can be pulled out through the buffeting region until an accelerated stall is produced.

(1) Do not enter a dive with the airplane trimmed nose heavy. Trim for approximately 275 mph IAS, and start into the dive. It probably will not be necessary to use the trim tabs for recovery, although the stick forces will be fairly high at high speeds. In any case, use the trim tab slowly and cautiously.

(2) If buffeting occurs during the dive, begin a gradual pullup immediately. The buffeting may

become more severe, but the airplane can be pulled out to about 2.5 gs at 30,000 feet, and to increased gs at lower altitudes before it stalls.

(3) The accelerated stall is similar to the normal low-speed stall, excepting that buffeting may occur before the stall is reached and the pitching motion, in the stall, is considerably more violent.

(4) If external load items are carried, use caution in making high-speed dives and pullouts. (See paragraph 1 *a* (1).) In all cases, avoid violent movement of the controls if the airplane is buffeting.

*c.* **DIVING LIMITATIONS.** — See section II, paragraph 1*a*(2)*(g)*.

**WARNING**

Above 30,000 feet extreme caution must be exercised in dives to avoid serious compressibility effects.

**17. NIGHT FLYING.**

*a.* The position lights are controlled by an "ON," "OFF," "DIM" toggle switch on the main electrical panel. The recognition lights are controlled by an individual switch box mounted on the right rail in the pilot's compartment. (See 1, figure 8.) Desired color combinations can be selected and "keyed" by the button switch adjacent to the switches.

**Note**

The position lights cannot be seen from within the airplane.

*b.* A retractable landing light is installed in the lower surface of each outer wing panel. Two lights, mounted on either side of the nose gear, are for use while taxiing. The control switches are on the pilot's electrical panel.

**Note**

Limit the use of the recognition and landing lights as much as possible. Switch off the landing lights and use the taxiing lights as soon as possible after landing.

**18. APPROACH AND LANDING.**

*a.* **ENGINE AND FLIGHT CONTROL SETTINGS.**

(1) **FUEL SELECTOR VALVE CONTROLS:** Fullest main tanks. Make *certain* that the fullest main tanks have been selected. Do not land while using fuel from the drop tanks.

(2) **MIXTURE CONTROL:** "AUTO RICH."

(3) **AUTOMATIC PILOT:** "OFF."

(4) **DE-ICER:** "OFF."

(5) **TURRET:** Check to be sure guns are stowed in the forward position.

(6) **BOOSTER PUMPS:** "HIGH."

(7) **COMPARTMENT HEATERS:** "OFF."

(8) **PROPELLER SWITCHES:** "CONSTANT SPEED."

(9) **PROPELLER GOVERNOR CONTROLS:** Increase to 2400 rpm on the base leg.

(10) **CROSS FEED:** "OFF.'

(11) **LANDING GEAR:** Lower at less than 175 mph IAS. Be sure it is down and locked. Check the warning horn and the indicator on the instrument panel. Also, look to see that the wheels are extended. As the landing gear descends the nose of the airplane rises until the wheels are about half way down. From that point on, the airplane's nose drops and it is necessary to re-trim the elevator.

(12) **COWL FLAPS:** Normally, closed for landing.

(13) **HYDRAULIC AND ACCUMULATOR PRESSURE:** 850-1000 psi.

(14) **FLAPS:** "DOWN," below 175 mph IAS.

*b.* **RECOMMENDED APPROACH SPEEDS.**—The speeds given below are for use just prior to reaching the field ("over the fence" speeds) with flaps and gear down.

(1) **POWER ON:** 105-110 mph IAS.

(2) **POWER OFF:** 115 mph IAS.

*c.* **POWER-OFF LANDING.**

(1) Power-off landings require an unusually steep approach to maintain gliding speed. This is because of the exceptionally large flap area on the airplane.

(2) An airspeed of 130-140 mph IAS should be maintained during the approach and the flare out should be started, smoothly and gradually, at about 100 feet above the runway.

(3) In power-off landings one-third to full flaps may be used. Naturally, the greater the angle of flap, the steeper the glide.

(4) Always close the cowl flaps in making a power-off landing and avoid letting the engines cool below 100°C. When the throttles are opened to correct the glide open them smoothly and slowly.

**CAUTION**

The rate of descent is very much higher without power than with power. Also the airplane lands at a slightly higher speed.

*d.* **MINIMUM RUN LANDING.**

(1) **SHORT FIELD LANDING.**

*(a)* Establish a normal full flap glide with power, in order to undershoot the field slightly.

*(b)* Hold normal speed throughout the flare out, then gradually pull up the nose, increase power, and go directly into slow flying. Do this when close enough to the ground to land as soon as power is cut.

*(c)* As the landing point is approached, cut the power. Since the airplane is already at the power-off stalling speed, the landing will be made readily. Allow the nose wheel to touch the ground as soon as practicable and use maximum permissible brakes, but do not skid the tires.

*e.* **TAKE-OFF IF LANDING IS NOT COMPLETED.**

(1) Level off, advance the throttles to climbing manifold pressure, and raise the landing gear lever immediately.

(2) Advance the throttles and propeller controls to take-off power.

(3) Retrim the elevator trim tabs for level flight.

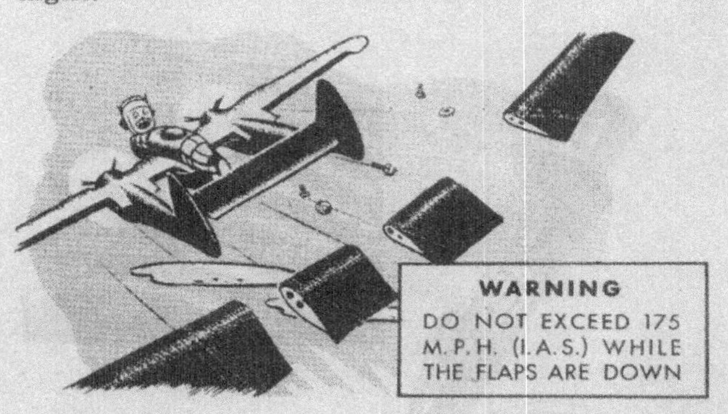


(4) Retract the flaps to one-third (20 degrees) as soon as possible after the gear has locked up.

(5) Proceed as in a normal, partial-flap take-off.

**19. STOPPING ENGINES.**

*a.* **PROCEDURE.**

(1) Open the cowl flaps fully.

(2) If the cylinder head temperatures are high, idle the engines until they fall below 200°C.

(3) Leave the propeller controls in the "INC. RPM" position.

(4) Stop each engine by moving the mixture controls to "IDLE CUT-OFF" and, simultaneously, opening the throttles fully.

(5) When the engines have stopped, turn all ignition and battery switches "OFF."

(6) Leave the mixture controls in "IDLE CUT-OFF."

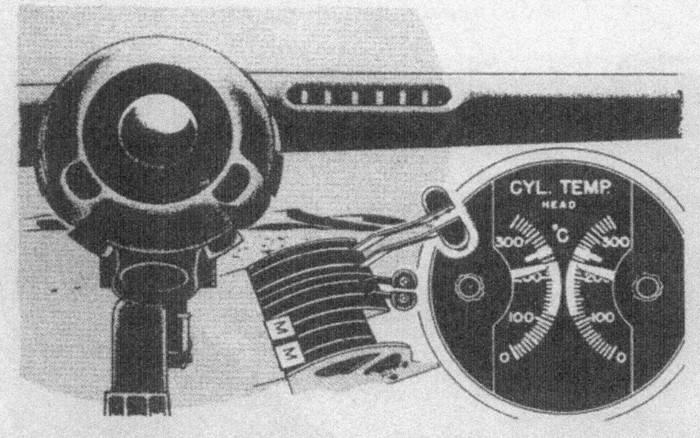


*b.* **OIL DILUTION.**—If a cold-weather start is expected, use the oil dilution system before the engines are stopped and after the cylinder-head temperatures have dropped below 205°C.

(1) Operate the engines 1000-1200 rpm.

(2) Turn the fuel booster pumps "OFF."

(3) Maintain oil temperatures between 5°C and 50°C.

(4) Dilute the oil for the following time periods, in accordance with the anticipated temperature, by placing the oil dilution switches in the "ON" position.

+4°C (+40°F) to −12°C (+10°F) 3 minutes
−12°C (+10°F) to −29°C (−20°F) 6 minutes
−29°C (−20°F) to −46°C (−50°F) 9 minutes

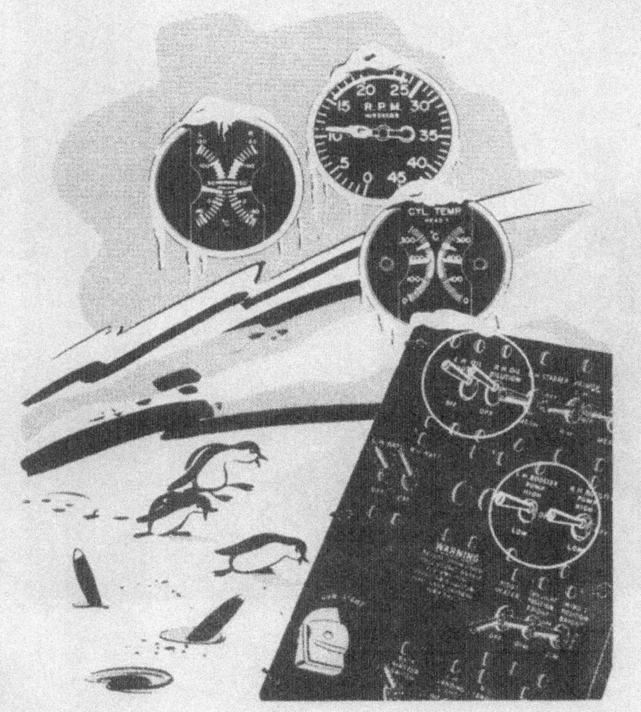


**CAUTION**

If oil temperatures cannot be maintained below 50°C during the dilution period, stop the engines, release the oil dilution switches, and allow the oil to cool well below 40°C. When cooled sufficiently, restart the engines, dilute the remainder of the time period, stop the engines, and release the dilution switches. It may be necessary to repeat this procedure several times in order to maintain oil temperatures below +50°C. The total time that the dilution switches are engaged should, nevertheless, conform to the time periods for the anticipated ground temperatures as charted under paragraph (4), immediately preceding.

(5) Add one minute dilution time for each additional 5°C (9°F) below −50°C (−58°F).

(6) In extremely cold weather, adequate dilution will prevent oil cooler or oil line failures due to high pressure developed by the scavenge pumps when the engine starts. However, for conservative operation, it is desirable to heat the oil lines, the oil cooler, and the accessories at the same time the portion of the engine forward of the diaphragm is being treated.

## 20. BEFORE LEAVING THE PILOT'S COMPARTMENT.

*a.* **ALL SWITCHES:** "OFF."

*b.* **FUEL SELECTOR VALVE CONTROLS:** Turned to fullest main tanks.

*c.* **PARKING BRAKES:** "ON."

**CAUTION**

Do not use the parking brakes when they are hot or during cold-weather operation. Use wheel chocks.

*d.* **ENGINE COWL FLAPS:** "OPEN."

*e.* **FLIGHT CONTROL LOCK:** "ON." Lock the flight controls by moving the lever into the "ON" position. Gently move the aileron, elevator, and rudder controls in the neutral position until they lock. The throttles must be "CLOSED."

*f.* Upon leaving the airplane, close and lock the entrance doors. Leave the windows slightly open to permit the circulation of air.

**AIRSPEED INSTALLATION CORRECTION TABLE**

| IAS | Correction |
|---|---|
| **FLAPS RETRACTED** | |
| 100 | Subtract 1 mile |
| 120 | Subtract 2 miles |
| 140 | Subtract 2 miles |
| 165 | Subtract 3 miles |
| 185 | Subtract 3 miles |
| 205 | Subtract 4 miles |
| 225 | Subtract 4 miles |
| 245 | Subtract 5 miles |
| 265 | Subtract 5 miles |
| 285 | Subtract 5 miles |
| 305 | Subtract 6 miles |
| 360 | Subtract 8 miles |
| 410 | Subtract 10 miles |
| **FLAPS AND GEAR EXTENDED** | |
| 100 | Add 4 miles |
| 120 | Add 5 miles |
| 140 | Add 6 miles |

***Figure 30 — Airspeed Installation Correction Table***

| | |
|---|---|
| 1. WATER TANK HEATERS | 20 AMPS. |
| 2. OIL COOLER FLAP MOTORS | 20 AMPS. |
| 3. GUN TURRET — STANDBY | 90 AMPS. |
| — OPERATING | 170 AMPS. |
| 4. CANNON FIRING SOLENOIDS | 40 AMPS. |
| 5. BOMB RELEASE SOLENOIDS | 40 AMPS. |
| 6. FUEL BOOSTER PUMPS | EACH, 7.5 AMPS. |
| 7. LANDING LIGHTS | 44 AMPS. |
| 8. RADAR EQUIPMENT | 101 AMPS. |
| 9. A.C. INVERTER OR RADIO EQUIPMENT | 100 AMPS. |

**Note**

The generators in this airplane have a maximum rated output of 300 amperes each. However, for emergency operation, they can be operated at 400 amperes for fifteen minutes or 500 amperes for eight minutes. (Operation above 300 amperes is beyond specification requirements and therefore the life of the generator cannot be guaranteed under these conditions.) For single engine operation, watch the ammeter on the active generator and do not operate so much equipment that the above noted ratings are exceeded.

***Figure 31—Electrical and Radio Equipment with Large Current Requirements***

# POWER PLANT CHART

**AIRCRAFT MODEL(S)**
**P-61C**

**PROPELLER(S)**
**(2) CURTISS C642 S-B (SPA-9)**

**CARBURETOR MODEL**
**(2) PR-58E2-2**

**ENGINE MODEL(S)**
**(2) R-2800-73**

| GAUGE READING | FUEL PRESS. | OIL PRESS. | OIL TEMP. | CYL. HD. TEMP. | WATER PRESS. (W.E.P.) |
|---|---|---|---|---|---|
| DESIRED | 18 | 75-95 | | 130 - 232 | 18 |
| MAXIMUM | 19 | 110 | 85 [3] | 260 | 19 |
| MINIMUM | 17 | 50 | 40 | 100 | 17 |
| IDLING | 11 | 25 | | | |

MAXIMUM PERMISSABLE DIVING RPM: **3120**
MINIMUM RECOMMENDED CRUISE RPM: **1400**
MAXIMUM RECOMMENDED TURBO RPM: **22,000**

OIL GRADE: (S) **1120** (W) **1100**
FUEL GRADE: **AN-F-28-2, 100/130**

| OPERATING CONDITION | WAR EMERGENCY (COMBAT EMERGENCY) | MILITARY POWER (NON-COMBAT EMERGENCY) | NORMAL RATED (MAXIMUM CONTINUOUS) | MAXIMUM CRUISE [4] (NORMAL OPERATION) |
|---|---|---|---|---|
| TIME LIMIT | 5 MINUTES | 15 MINUTES | UNLIMITED | UNLIMITED |
| MAX CYL HD TEMP. | 260 | 260 | 260 | 232 |
| MIXTURE | AUTO-RICH | AUTO-RICH | AUTO-RICH | AUTO-LEAN |
| RPM | 2800 | 2800 | 2600 | 2600 |

| MANIF PRESS. | SUPER CHARGER | FUEL [1] *Gal/Min* | MANIF PRESS. | SUPER CHARGER | FUEL [1] *Gal/Min* | STD TEMP. °C | PRESSURE ALTITUDE | STD TEMP. °F | MANIF PRESS. | SUPER CHARGER | FUEL GPH [2] | MANIF PRESS. | SUPER CHARGER | FUEL GPH [2] |
|---|---|---|---|---|---|---|---|---|---|---|---|---|---|---|
| | | | | | | -55.0 | 40,000 FT | -67.0 | | | | | | |
| | | | | | | -55.0 | 38,000 FT | -67.0 | | | | | | |
| | | | | | | -55.0 | 36,000 FT | -67.0 | | | | | | |
| | | | | | | -52.4 | 34,000 FT | -62.3 | | | | | | |
| | | | | | | -48.4 | 32,000 FT | -55.1 | | | | | | |
| | | | | | | -44.4 | 30,000 FT | -48.0 | | | | | | |
| | | | | | | -40.5 | 28,000 FT | -40.9 | | | | | | |
| | | | | | | -36.5 | 26,000 FT | -33.7 | | | | | | |
| | | | | | | -32.5 | 24,000 FT | -26.5 | | | | | | |
| | | | | | | -28.6 | 22,000 FT | -19.4 | | | | | | |
| | AUTOMATIC | | | AUTOMATIC | | -24.6 | 20,000 FT | -12.3 | | AUTOMATIC | | | AUTOMATIC | |
| | | | | | | -20.7 | 18,000 FT | - 5.2 | | | | | | |
| | | | | | | -16.7 | 16,000 FT | 2.0 | | | | | | |
| | | | | | | -12.7 | 14,000 FT | 9.1 | | | | | | |
| | | | | | | - 8.8 | 12,000 FT | 16.2 | | | | | | |
| | | | | | | - 4.8 | 10,000 FT | 23.4 | | | | | | |
| | | | | | | - 0.8 | 8,000 FT | 30.5 | | | | | | |
| | | | | | | 3.1 | 6,000 FT | 37.6 | | | | | | |
| | | | | | | 7.1 | 4,000 FT | 44.7 | | | | | | |
| | | | | | | 11.0 | 2,000 FT | 51.8 | | | | | | |
| | | | | | | 15.0 | SEA LEVEL | 59.0 | | | | | | |

**GENERAL NOTES**

[1] *Gal/Min*: APPROXIMATE US GALLON PER *MINUTE* PER ENGINE
[2] GPH: APPROXIMATE US GALLON PER HOUR PER ENGINE
FT: MEANS FULL THROTTLE OPERATION.
VALUES ARE FOR LEVEL FLIGHT WITH RAM.

FOR COMPLETE CRUISING DATA SEE APPENDIX I
NOTE: TO DETERMINE CONSUMPTION IN BRITISH IMPERIAL UNITS, MULTIPLY BY 10 THEN DIVIDE BY 12. RED FIGURES ARE PRELIMINARY SUBJECT TO REVISION AFTER FLIGHT CHECK.

**TAKE-OFF CONDITIONS:**
2800 RPM, 54.0 IN. Hg, "AUTO-RICH"

**CONDITIONS TO AVOID:**
TURBO PULSATION (SEE FIGURE 26)

**SPECIAL NOTES**

[3] 98°C ALLOWED DURING CLIMB
[4] "AUTO LEAN" MIXTURE MAY BE USED FOR OPERATION AT POWERS UP TO RATED POWER PROVIDING CYLINDER HEAD TEMPERATURE DOES NOT EXCEED 38°C. IF EITHER OF THESE LIMITS ARE EXCEEDED, "AUTO-RICH" MIXTURE MUST BE USED.

DATA AS OF 6-15-45 BASED ON ENGINE DATA AND PRELIMINARY FLIGHT TEST

AAFMC-526
4-1-44

*Figure 32 — Power Plant Chart*

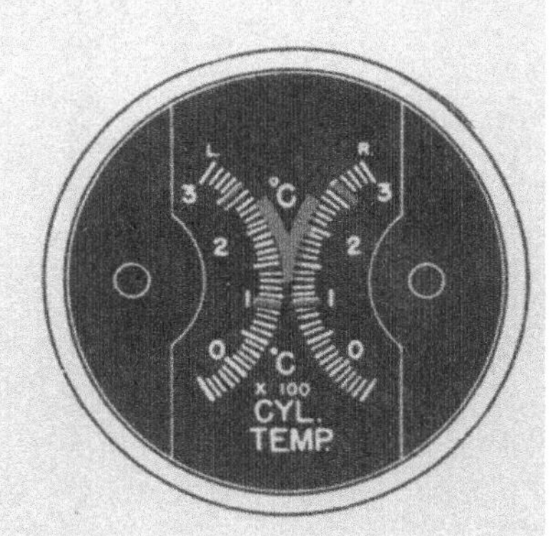

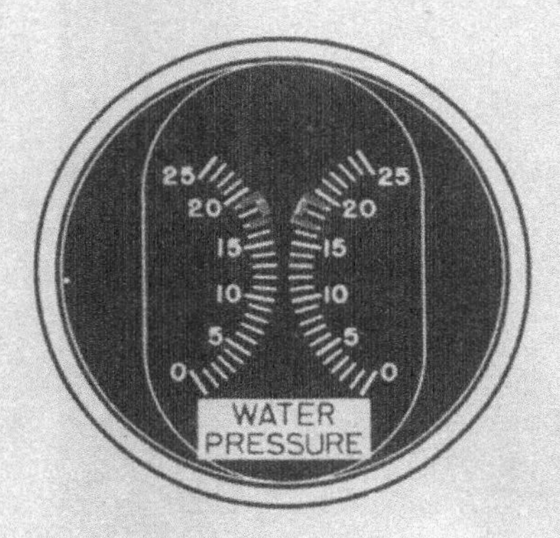


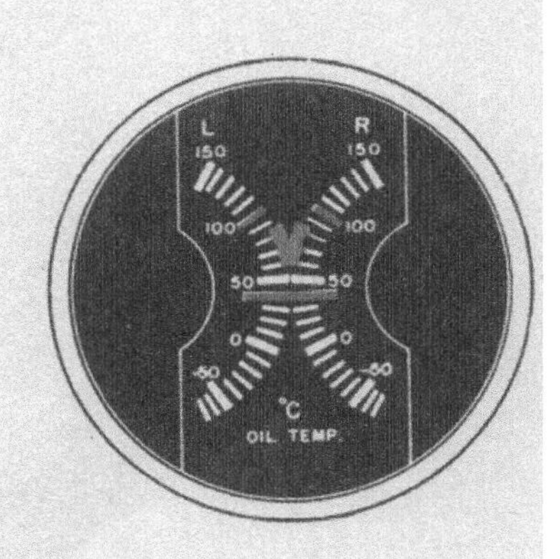


THE COLORS USED IN THE INSTRUMENT MARKINGS AND THEIR MEANINGS ARE:

GREEN: DESIRABLE OPERATING RANGE

BLUE: CRUISING RANGE

YELLOW: MAXIMUM RANGE

RED: LIMIT OR DANGER REGION

POWER LIMITS ARE FOR GRADE 100/30 FUEL

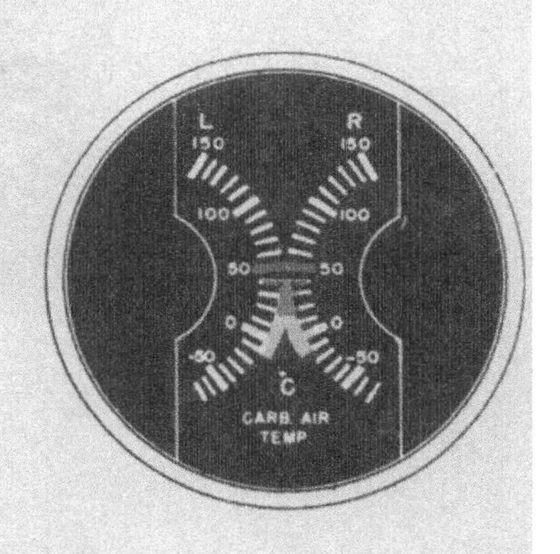

***Figure 33 — Instrument Range Markings***

239728

# SECTION IV
# EMERGENCY OPERATIONS

## 1. EMERGENCY EGRESS.

*a.* EXITS.—Two methods are provided whereby the pilot and gunner can leave the airplane in flight.

### WARNING

The pilot's hinged canopy is not intended as a bailout exit and it is not safe to use it as such, since it is forward of the propellers.

(1) QUICK BAILOUT.—In a quick bailout, first reduce the airspeed of the airplane as much as possible. This may be accomplished by pulling the airplane into a medium climb which has the dual advantage of gaining altitude and losing speed.

*(a)* The quickest exit can normally be made through the gunner's escape hatch. The hatch, which is the right-hand portion of the enclosure aft of the gunner's windshield, is released to fly away by turning the emergency release handle and pushing sharply.

*(b)* The gunner's seat is opposite the emergency exit.

*(c)* To reach the emergency exit, the pilot should tilt his seat backward by pressing the lever on the back of the seat behind his right shoulder. He then turns around on his knees and crawls back to the gunner's position. If the gunner's armor plate is locked in position, release it by pressing the spring-loaded handle at the bottom center, slide the bolt to the right, and swing the armor plate out of the way.

*(d)* Both pilot and gunner should climb through the escape hatch, head first, onto the inner wing. If conditions permit, they should slide head first and face down over the trailing edge of the inner wing.

(2) CONTROLLED BAILOUT. — If conditions permit control of the airplane, the safest way to bail out is through the entrance hatch in the nose wheel well.

*(a)* Lower the landing gear; then, if possible, put the airplane on automatic pilot, and fly at about 30 mph IAS above the stalling speed. Set the controls for a descent of about 1000 feet per minute. The gunner releases the entrance door by turning the door handle clockwise and, at the same time, pulling sharply on the red emergency release handle, located on the left-hand rail between the pilot and gunner. If the door does not drop immediately, stamp on it.

*(b)* The gunner should leave the airplane by leaning forward and diving out.

*(c)* The pilot tilts his seat backward as in a quick bailout. If it is possible to do so, he should step over the floor opening to the gunner's position, turn around to face the nose of the airplane, and dive through the opening. Otherwise, he should dive out head-first from the pilot's position.

(3) **RADIO OPERATOR'S EXIT.**—In either a quick or controlled bailout, the radio operator leaves the airplane through the rear entrance door.

*(a)* To release the rear entrance door, pull sharply on the red emergency release handle, located on the left side of the compartment and, at the same time, pull the normal release. This removes the hinge pins and, when it is given a sharp kick, the door drops free.

*(b)* Facing the rear of the airplane, the radio operator should jump out, feet first, with his hands raised before his face. As he drops and the lower part of his body is blown back in the slip stream, he shoves against the rear edge of the opening to prevent his head from hitting the crew nacelle.

*b.* **ALARM BELLS.** — There are two alarm bells; one beneath the radio operator's table on the right side of the compartment, and the other on the left side of the crew nacelle near the gunner's seat.

(1) The alarm bell operating switch is located on the left rail between the gunner's and pilot's positions. (See figure 34.)

(2) As soon as the necessity for bailing out is determined, the pilot should warn the gunner and radio operator by interphone, if possible. Also, he should ring the alarm bell three times, then leave it ringing.

**2. FIRE.**

*a.* **USE OF FIRE FIGHTING EQUIPMENT.**

(1) **HAND-OPERATED FIRE EXTINGUISHERS.**

*(a)* **DESCRIPTION.** — A carbon dioxide, hand-operated fire extinguisher is clipped to the bulkhead of the aft end of the gunner's compartment near the left-hand rail. A second, hand-operated, carbon dioxide fire extinguisher is located in the radio operator's compartment. A hand-operated, carbon tetrachloride fire extinguisher is installed in each main gear wheel well. An access door, marked "FIRE EXTINGUISHER," in the outboard side of each wheel well provides for the removal of the extinguisher.

(2) **ENGINE FIRE EXTINGUISHING SYSTEM.** (See figure 35.)

*(a)* **DESCRIPTION.** — A methyl bromide fire extinguishing system is installed in each engine nacelle for the purpose of combating engine section fires either in the air or on the ground. Switches for the system are mounted below the left cockpit rail aft of the propeller control panel. (See 2, figure 7.) Indicator lights are installed in the main instrument panel. (See figure 15.)

*b.* **ENGINE FIRES.**—At the first sign of a fire, if conditions permit, use the following procedure *on the affected engine.*

(1) **COWL FLAPS:** "OPEN."

(2) **FUEL SELECTOR VALVE:** "OFF."

(3) **PROPELLER:** Feathered.

(4) **IGNITION:** "OFF."

(5) **FIRE EXTINGUISHER SWITCH:** "ON."

(6) **DO NOT START THE ENGINE AGAIN.**

*c.* **WING FIRES.**—In the event of a wing fire, turn "OFF" all switches for landing, position, and recognition lights. Attempt to put out the fire by sideslipping the airplane.

*Figure 34 — Alarm Bells Switch*

*d.* **FUSELAGE FIRES.**—If a fire occurs in the crew nacelle, close all windows and ventilators. If the fire is in the electrical system, turn the master switch "OFF." If the blaze has started in a leaking fuel or hydraulic line, shut off the valves. Use the fire extinguisher immediately.

**3. ENGINE FAILURE DURING FLIGHT.**

*a.* **SINGLE ENGINE OPERATION.**—With one engine dead and its propeller windmilling, with military power on the other engine, and with rudder trim neutral, the airplane can be controlled down to its normal stalling speed of 118 mph IAS (gear and flaps up). Under normal conditions it is not necessary to cut the power on the good engine. Just change the mixture control to "AUTO-RICH." However, on the dead engine, feather as instructed in section II, paragraph 9 *b* (3).

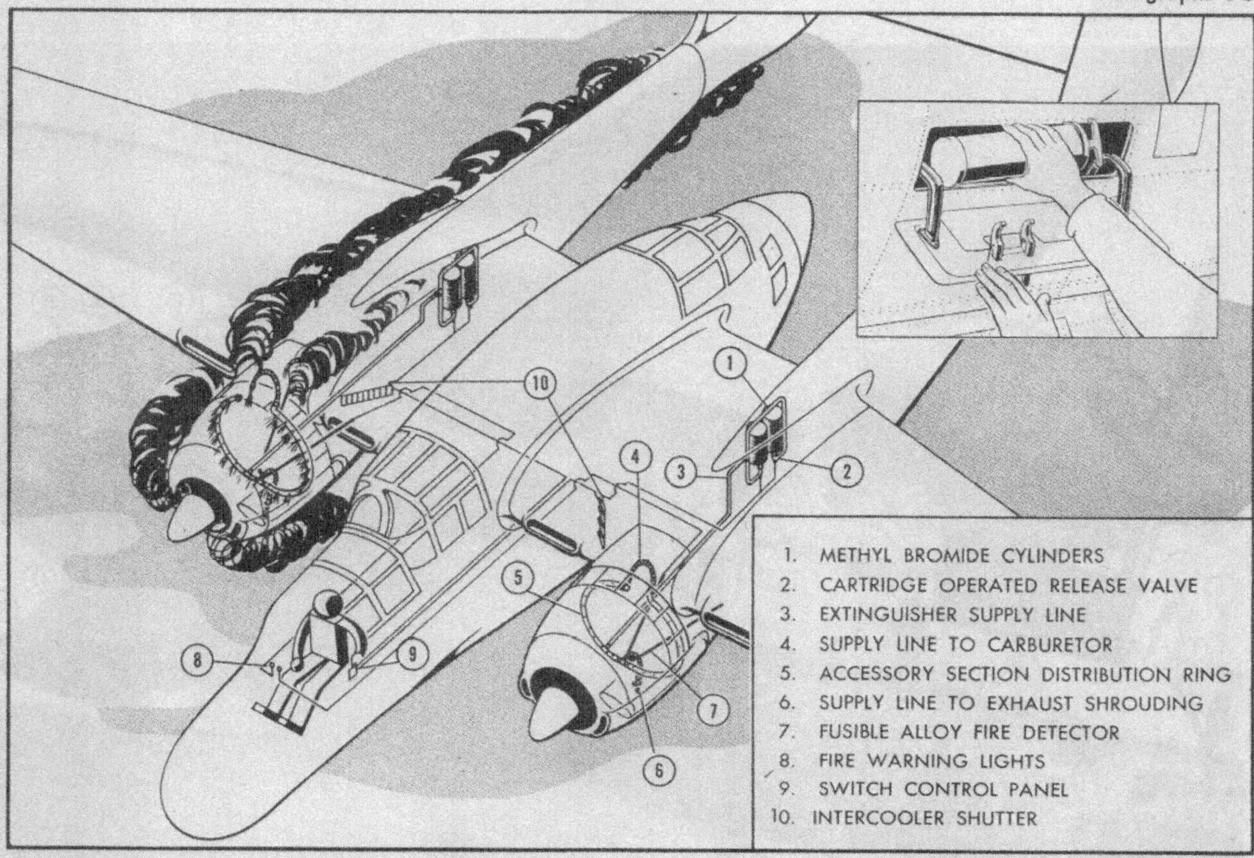


***Figure 35 — Engine Fire Extinguishing System***

*b.* SINGLE ENGINE LANDING.

(1) Make traffic pattern turns in the direction of the good engine, if practicable. Trim the airplane for straight and level flight, leaving a little rudder pressure on the side of the good engine. Be sure to keep the airspeed constant. Lose a little altitude on turns, if necessary, in order to maintain critical single engine speed or higher. It is desirable to maintain slightly more speed than is required for a two-engine approach.

(2) Lower the landing gear only after turning onto the approach leg and when sure of making the field.

(3) Lower the flaps not more than one-half until certain of reaching the field.

(4) The decision to go around again because of possible over-shooting should be made at a time when the airplane still has a safe margin of altitude and speed.

(5) If it is necessary to go around again, raise the landing gear immediately to reduce drag. Apply power gradually, raise the flaps SLOWLY to 20 degrees (one-third) as soon as speed permits.

**4. BOMB EMERGENCY RELEASE.**

*a.* A manual emergency release for bombs or drop tanks is mounted on the compartment floor to the left of the pilot. To operate, pull the T-handle up. This jettisons the drop installations on all four shackles. Bombs are dropped safe. (See figure 53.)

**5. HYDRAULIC SYSTEM FAILURE.**

*a.* In case of hydraulic system failure, all hydraulic equipment except the automatic pilot may be operated by the hand pump for a short time unless the fluid leaks out faster than the hand pump can supply pressure. Usually, a hydraulic system failure will be indicated by a low pressure gage reading (850-1000 psi normal).

**Note**

The brakes and turret ejection chute doors operate from the accumulator. All other units operate directly from the main pressure system. Normally, it is advisable to save the hand-pump pressure for flap and brake operation.

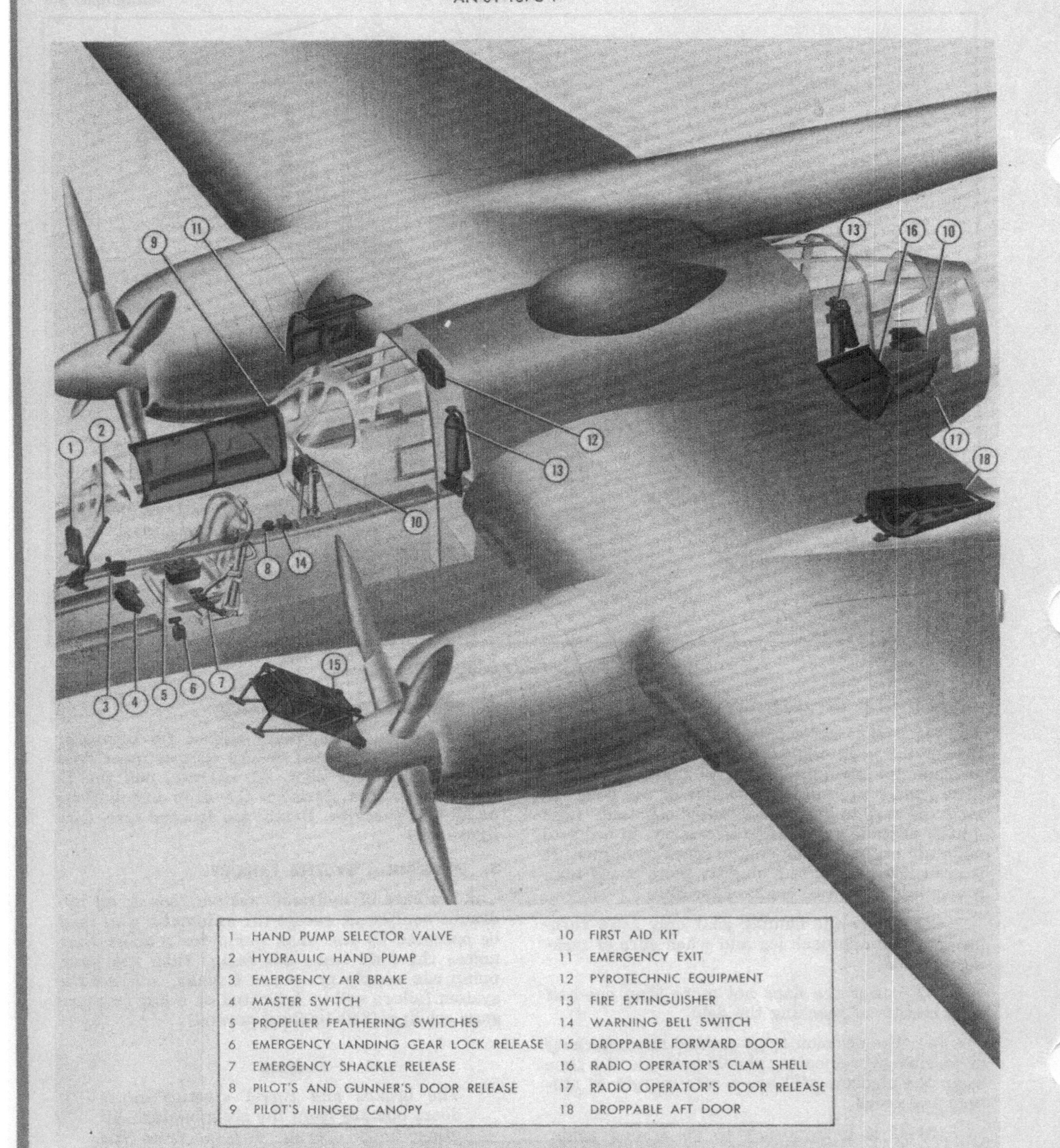


**Figure 36 — Emergency Equipment and Exits**

*b.* **HAND PUMP OPERATION.** (See figure 37.)

(1) Turn the hand-pump selector-valve control to "SYSTEM," "ACCUMULATOR," or "FIGHTER BRAKE AND LANDING FLAP EMERGENCY," as desired.

(2) On the appropriate selector valve, select the desired position for the unit to be operated. In the case of the fighter brakes, there is no selector valve; however, the wing flap selector valve must be in "NEUTRAL."

(3) Pump vigorously to produce the required pressure.

(4) When the operation is completed, return the hand-pump selector-valve control to "NEUTRAL."

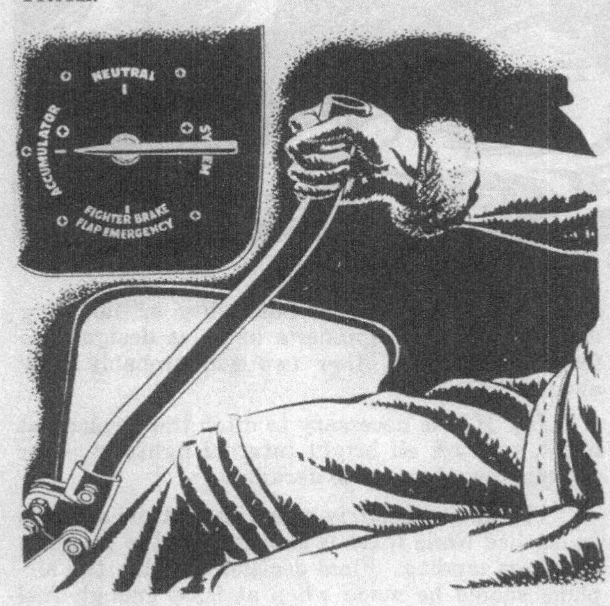


***Figure 37 — Hand Pump Operation***

## 6. LANDING GEAR EMERGENCY OPERATION.

*a.* EMERGENCY EXTENSION.

(1) If the landing gear cannot be operated hydraulically, place the selector-valve control in the "DOWN" position.

(2) Pull up the landing gear emergency release handle to extend the landing gear. (See 11, figure 7.)

(3) Hold the handle up until the nose gear is locked down, THEN RELEASE.

**Note**

When using the emergency landing gear release, reduce speed to 130 mph IAS or less. If the main gear fails to release, shake the airplane longitudinally.

*b.* EMERGENCY RETRACTION.

(1) Pull the emergency landing gear release handle up.

(2) The landing gear release, in this position, holds the down locks open. The main landing gear will retract upon contact with the ground. Because of its position, the nose gear will remain extended and the airplane will land on the nose wheel and booms.

## 7. BRAKE EMERGENCY OPERATION.

*a.* An emergency air brake system is provided for use in event of hydraulic system failure. The emergency air brake lever is located on the pilot's compartment rail towards the left front corner. (See 16, figure 6.) The storage bottle contains sufficient air for approximately four applications of the brakes.

## 8. MANIFOLD PRESSURE REGULATOR.

*a.* If the manifold pressure regulator fails because of a loss of oil pressure, manual control is effective up to approximately 35 inches Hg.

*b.* If the manifold pressure becomes dangerously high during a dive at medium or high power, use the normal procedure to reduce power. See section II, paragraph **11 *c* (2).**

## 9. LANDING WITH WHEELS RETRACTED.

If it is necessary to make a gear-up landing, observe the following procedure:

*a.* If time and other conditions permit, fire all 20-mm cannon ammunition toward an unpopulated area.

*b.* Turn off the master heater switches, located on the generator control panel, at least two min-

*Figure 38 — Abandoning the Airplane in the Water*

utes before landing so that the fuel in the lines and combustion chambers may be consumed.

*c.* Drain the airplane's oxygen supply completely by turning the red emergency knobs on the oxygen regulators to "ON."

*d.* Fully extend the wing flaps.

*e.* Make the approach just above stalling speed.

*f.* Pull the mixture controls back to "IDLE CUT-OFF."

*g.* Turn the master ignition switch and the master battery switch "OFF."

*h.* Land at the slowest practical speed.

**10. LANDING IN WATER (DITCHING).**

*a.* **PREPARATION FOR DITCHING.**

(1) If it becomes necessary to ditch the airplane, warn the crew by interphone or give the warning bell six short rings for standby and one long ring for ditching.

(2) If time permits, all ammunition in the airplane should be fired. If drop tanks or bombs are installed, they must be dropped before ditching the airplane.

(3) The radio operator and gunner must check to make sure that both fore and aft entrance doors are securely closed. The radio operator's overhead hatch, the escape hatch at the gunner's position, and the pilot's hinged canopy must be opened to eliminate the possibility of jamming upon impact. The gunner's hatch is designed to fly away and the other two will probably blow away.

(4) If it is necessary to ditch the airplane at night, turn off all bright internal lights in order to accustom the eyes to darkness.

(5) It is absolutely essential that the airplane be ditched while there is sufficient fuel to permit a power-on landing. Final decision to ditch the airplane should be made when at least enough fuel for 15 minutes' operation remains, unless the pilot is certain that he can reach land. Use of power assures adequate control and ability to flatten out the airplane for proper alighting on the water.

(6) Shortly before the airplane touches the water, each crew member should unfasten his parachute and free himself from the straps. BE CERTAIN THAT BOTH THE SAFETY BELT AND SHOULDER HARNESS ARE IN PLACE AND FASTENED.

(7) Stow both sighting stations in elevation and azimuth. Be sure the sighting arm is not in front of the crew member.

*b.* **DITCHING PROCEDURE.**

(1) Land at the slowest possible IAS. DO NOT LOWER THE LANDING GEAR. Wing flaps may be lowered as much as one-third (20 degrees) in order to reduce landing speed.

(2) In a crosswind-approach along a swell, ditch the airplane on the up-slope of the swell.

(3) In a steep swell, ditch the airplane along the top of the swell unless there is a very strong crosswind. In ditching across the swell, the airplane should be put down on an up-slope toward the top.

(4) The airplane should alight with the tail down. There will be a slight preliminary impact as the rear of the airplane strikes, followed by a severe impact and a rapid stop. The higher the rate of speed at which the airplane is landed, the more severe the impact, the greater the danger of structural collapse. If the landing is made too fast the airplane will tend to bounce, greatly increasing the danger of collapse.

*c.* **ABANDONING THE AIRPLANE.** (See figure 38.)

(1) As soon as the airplane stops, the crew members must release their safety belts and shoulder harness and climb out through the top escape hatches onto the inner wing. Take the parachute along.

(2) Inflate the Mae Wests. Pull the individual dinghies loose from the parachutes, inflate them, and wriggle into them. Salvage the parachutes, if possible, as they will be useful for sails, cover, extra lines, and to facilitate location by other aircraft.

## 11. MISCELLANEOUS EMERGENCY EQUIPMENT.

*a.* **FIRST AID KITS.**—Two first aid kits are provided. One is attached to the gunner's armor plate and the other is located to the left and aft of the radio operator's door.

*b.* **PYROTECHNIC EQUIPMENT.** — A type M-8 Very pistol, signal cartridges, and parachute flares are carried in the gunner's compartment in a container fixed to the bulkhead aft of the gunner's seat.

25631

**1. HEATING AND VENTILATING SYSTEM.**
(See figure 42.)

*a.* DESCRIPTION.—P-61C airplanes are equipped with two fuel-air type surface combustion heaters. The forward heater, mounted on the right-hand side of the bulkhead in the gunner's compartment, receives fuel from the right-hand engine and provides heat for the forward compartment. The aft heater, mounted below the radio operator's floor, receives fuel from the left-hand engine and provides heat for the radio operator's compartment and the guns. Ventilation also is provided through the heating system.

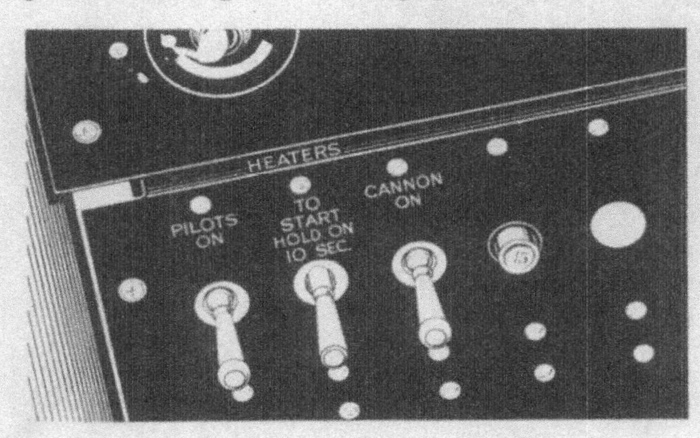


*Figure 39 — Forward Heater Switches*

*b.* CONTROLS.

(1) FORWARD HEATER.—The forward heater control switches, marked "HEATER," are located on the generator control panel. (See figure 39.) They consist of a master control switch with "ON" and "OFF" positions, and an energizing switch with an "OFF" position and an "ON" position, labeled 'TO START HOLD ON 10 SEC." A manual fuel pressure control, marked "CABIN HEAT," is installed on the pilot's compartment right rail, immediately above the generator control panel. (See figure 40.)

(2) AFT HEATER.—The aft heater master control switch, marked "CANNON," is located on the generator control panel. The aft heater energizing switch, marked "HEATER START," and the circuit breaker switch, marked "HEATER," are both located on the right crew nacelle rail opposite the radio operator's seat. A manual fuel-pressure override control, marked "PULL UP TO INCREASE TEMPERATURE," is mounted on the right crew nacelle rail opposite the radio operator. (See figure 41.)

*Figure 40 — Cabin Heat Control*

(3) AUTOMATIC SAFETY SWITCHES.—Three thermal safety switches, located on the output side of each heater, thermostatically shut off the heater if it becomes excessively hot.

(4) CABIN AIR OUTLET CONTROLS.—An outlet is provided at each of the three crew stations. Each outlet is equipped with a "CABIN

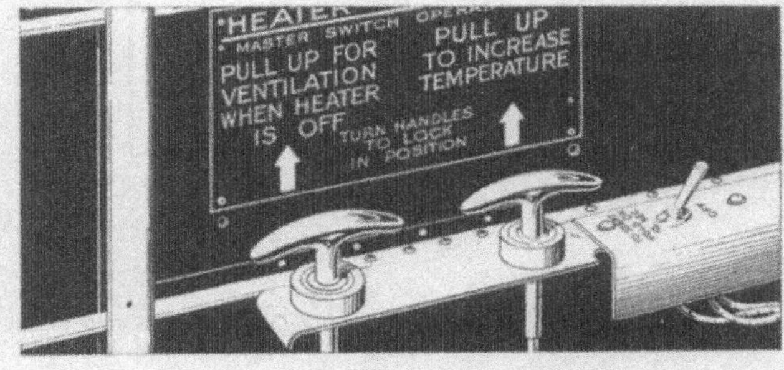


*Figure 41 — Aft Heater Controls*

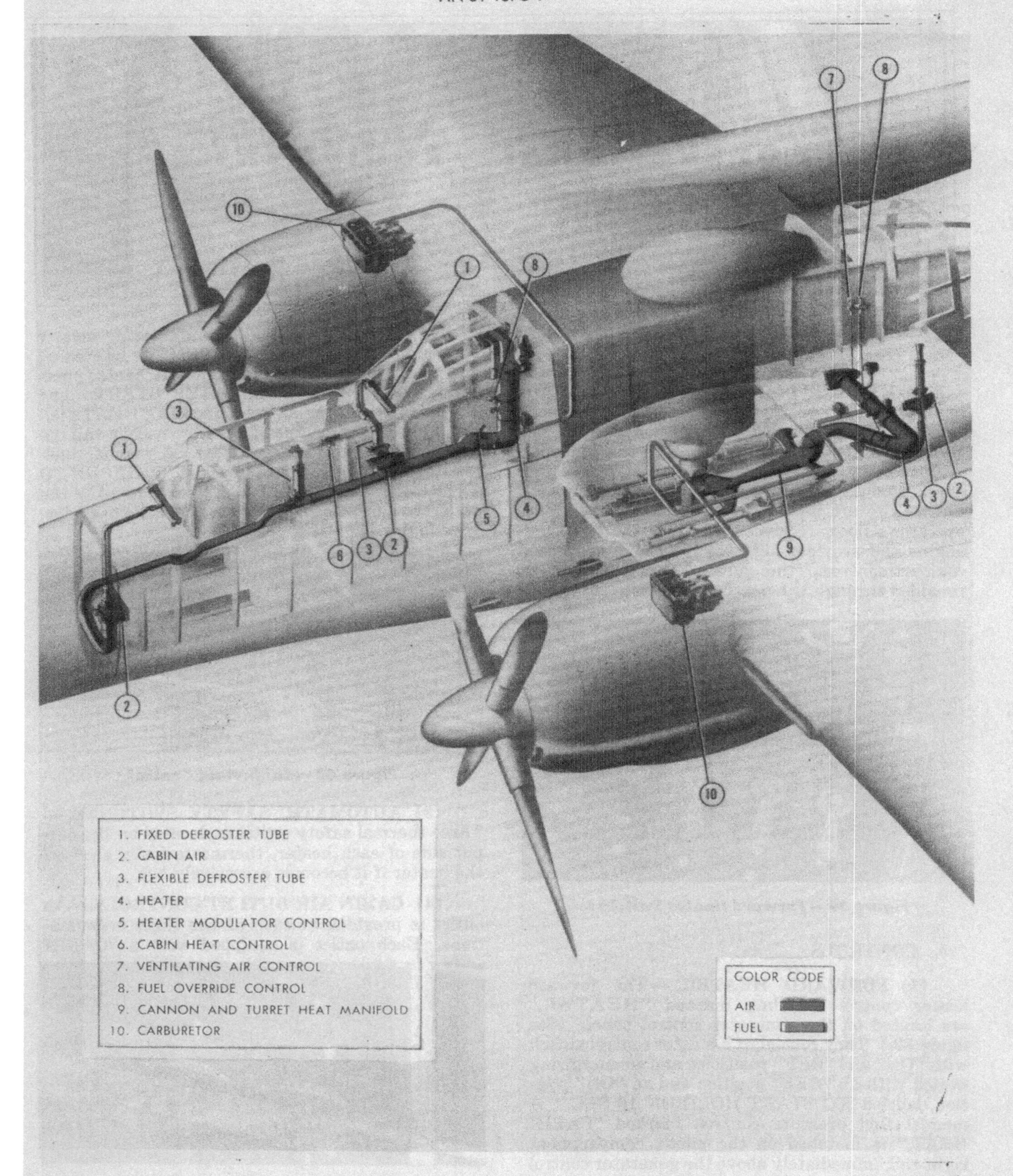


***Figure 42 — Heating and Ventilating System***

AIR" control having "OPEN" and "CLOSED" positions. (See figure 43.)

(5) **COLD AIR VENTILATION CONTROLS.** The cold air ventilation control for the forward heater is located on the floor at the left of the gunner's seat. The control for the aft heater is a T-handle on the gunner's compartment right rail, labeled "PULL UP FOR VENTILATION WHEN HEATER IS OFF."

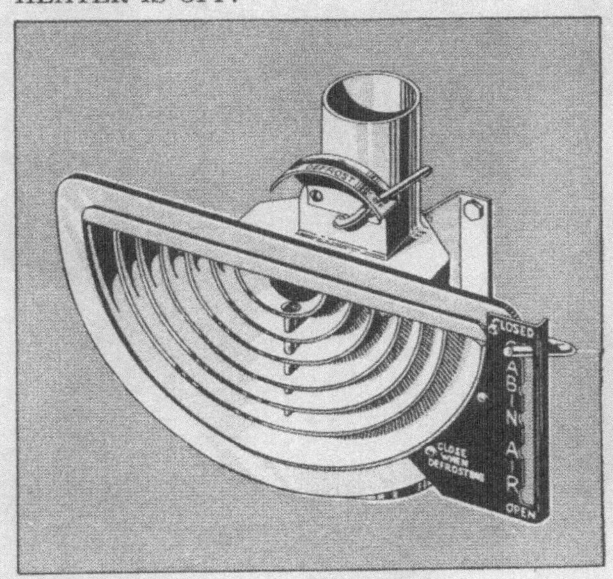


*Figure 43 — Cabin Air Outlet*

*c.* **OPERATION.**

(1) **FORWARD HEATER.**

*(a)* **TO START.**—Place the "FWD. CABIN" master-control switch in the "ON" position. Hold the energizing switch up against the spring tension to energize the fuel-solenoid shut-off valve. This should be accomplished in from two to 10 seconds. If heat is not noticed within 45 seconds after the switch is released, repeat the procedure.

*(b)* **TO INCREASE HEAT.**—To increase heat, move the manual control lever marked "CABIN HEAT" from the "ON" to the "WARMER" position.

*(c)* **TO STOP.**—Move the master control switch to the "OFF" position.

*(d)* **TO SECURE COLD AIR VENTILATION.**—Open the cold air ventilation control at the gunner's position. Place the "CABIN AIR" outlet control in the "OPEN" position.

**2. OXYGEN SYSTEM.** (See figure 45.)

*a.* **DESCRIPTION.**—The airplane is equipped with a low pressure oxygen system, operating at a working pressure of 400-450 psi.

*b.* **SOURCE OF SUPPLY.**—Oxygen is carried in 12 low-pressure oxygen cylinders, six in each engine nacelle, just aft of the wing trailing edge. The two upper cylinders in each engine nacelle provide oxygen to the pilot, the two middle cylinders to the radio operator, and the two bottom cylinders to the gunner.

*c.* **OXYGEN CONTROLS.** (See figure 44.)—Three demand type regulators are located, one to the right of each crew station. The emergency control is safety-wired in the "OFF" position. A pressure gage and a blinker flow indicator are mounted on a panel to the right of each crew station, adjacent to the regulator.

*Figure 44 — Oxygen Instruments*

**3. TURRET CONTROL SYSTEM.** (See figure 48.)

*a.* **GENERAL.**—A turret (see 16, figure 48), mounting four .50-caliber machine guns, is located between the gunner's and radio operator's compartments. The guns fire simultaneously. Each machine gun is provided with an ammunition case which will hold a maximum of approximately 560 rounds of ammunition. The turret rotates a full 360 degrees in azimuth and the guns move through an arc from 20 degrees below to 90 degrees above horizontal. The movement of the turret and guns is electrically controlled from two sighting stations (see 8 and 17, figure 48); one operated by the gunner and the other by the radio operator.

*b.* **CONTROLS.**

(1) **TURRET CONTROL BOX.** (See figure 46.)—The turret control box is mounted below the right rail in the gunner's compartment. It contains a switch controlling the amplidyne, dynamotor, and air compressor; and the "GUNS" switch which provides a means of disconnecting the firing circuits. The box also contains four circuit breakers (one of which is inoperative), a terminal post, and three AN connectors.

(2) **SIGHTING STATION.** (See figure 47.)—The sighting station firing controls are mounted

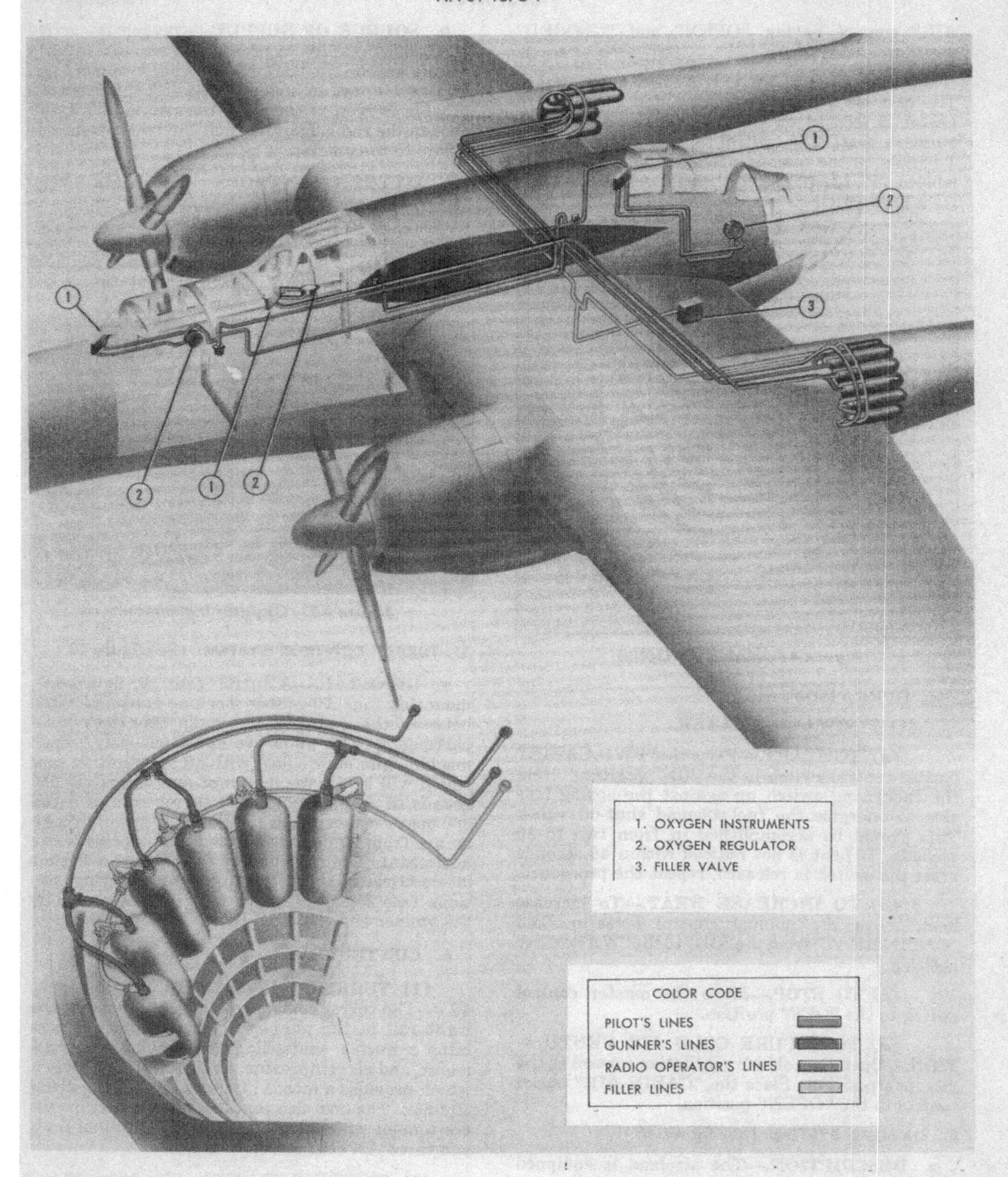


***Figure 45 — Oxygen System***

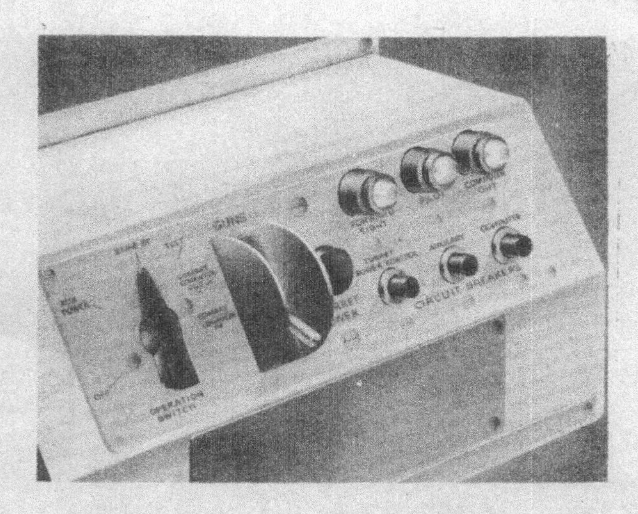

***Figure 46 — Turret Control Box***

on the two grips attached to the sighting arm. Each grip contains a palm action switch and a trigger switch. The trigger switch is mounted on the side of the grip away from the operator. All four turret guns are fired by the trigger on either sighting station. The palm action switches are mounted on the outboard side of the grips. They are depressed by the palm of the hand and pressure on either or both provides complete control over the turret. Movement of the sighting station in azimuth rotates the turret in azimuth, and elevation of the sighting arm elevates the turret guns. (The button switch on the left grip provides interphone communication.)

(3) **PILOT'S CONTROLS.**—The pilot's control consists of a firing trigger on the right forward side of the control wheel.

c. **OPERATION OF THE TURRET FIRE CONTROL SYSTEM.**

(1) Immediately after take-off, place the operation switch in the "STANDBY" position and leave it there until just before landing. This permits the system to warm up and the air compressor to build up pressure, insuring proper operation of the guns throughout the entire period of the mission.

(2) Unstow the sighting station in elevation by pulling out the elevation stowage pin and lower-

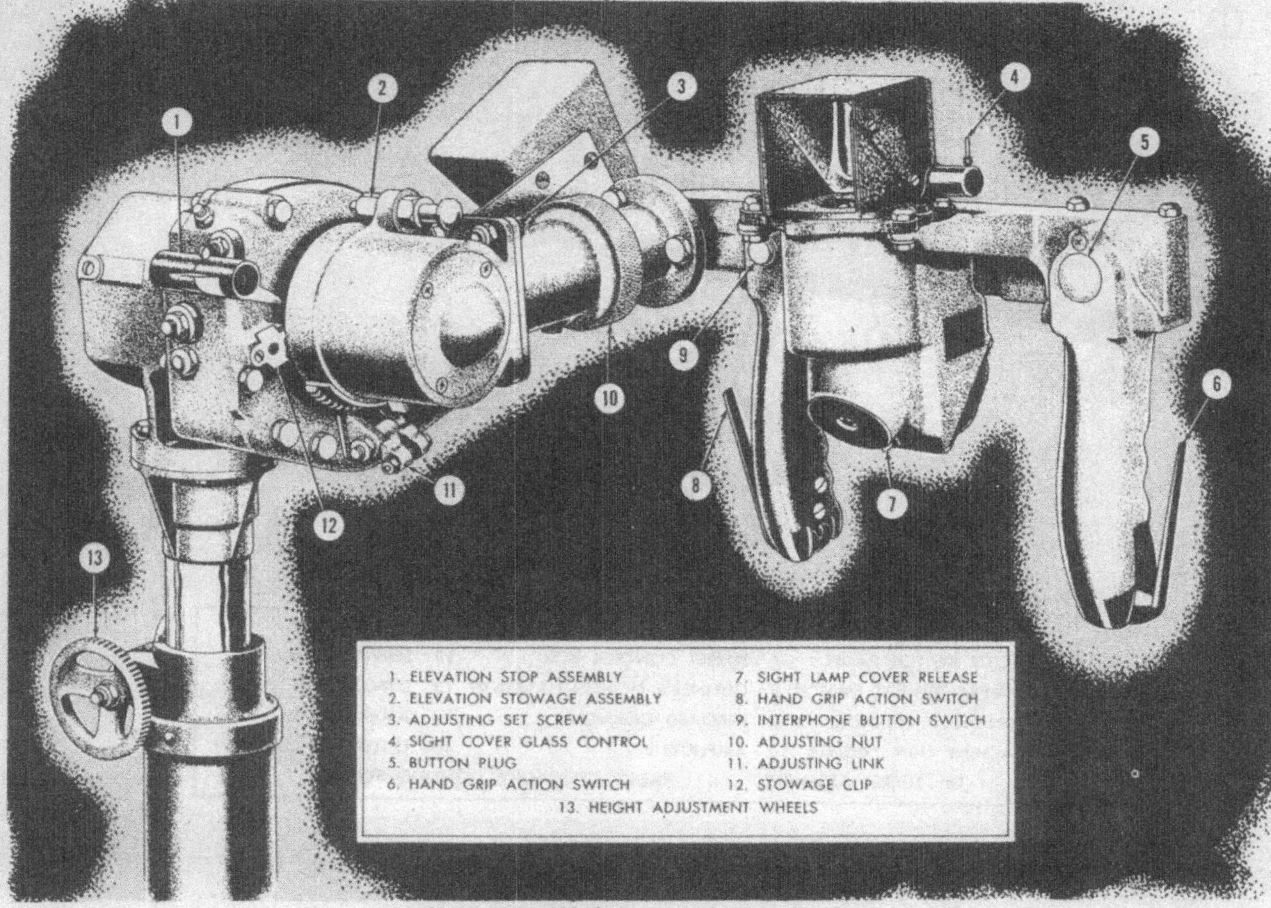


***Figure 47 — Sighting Station Controls***

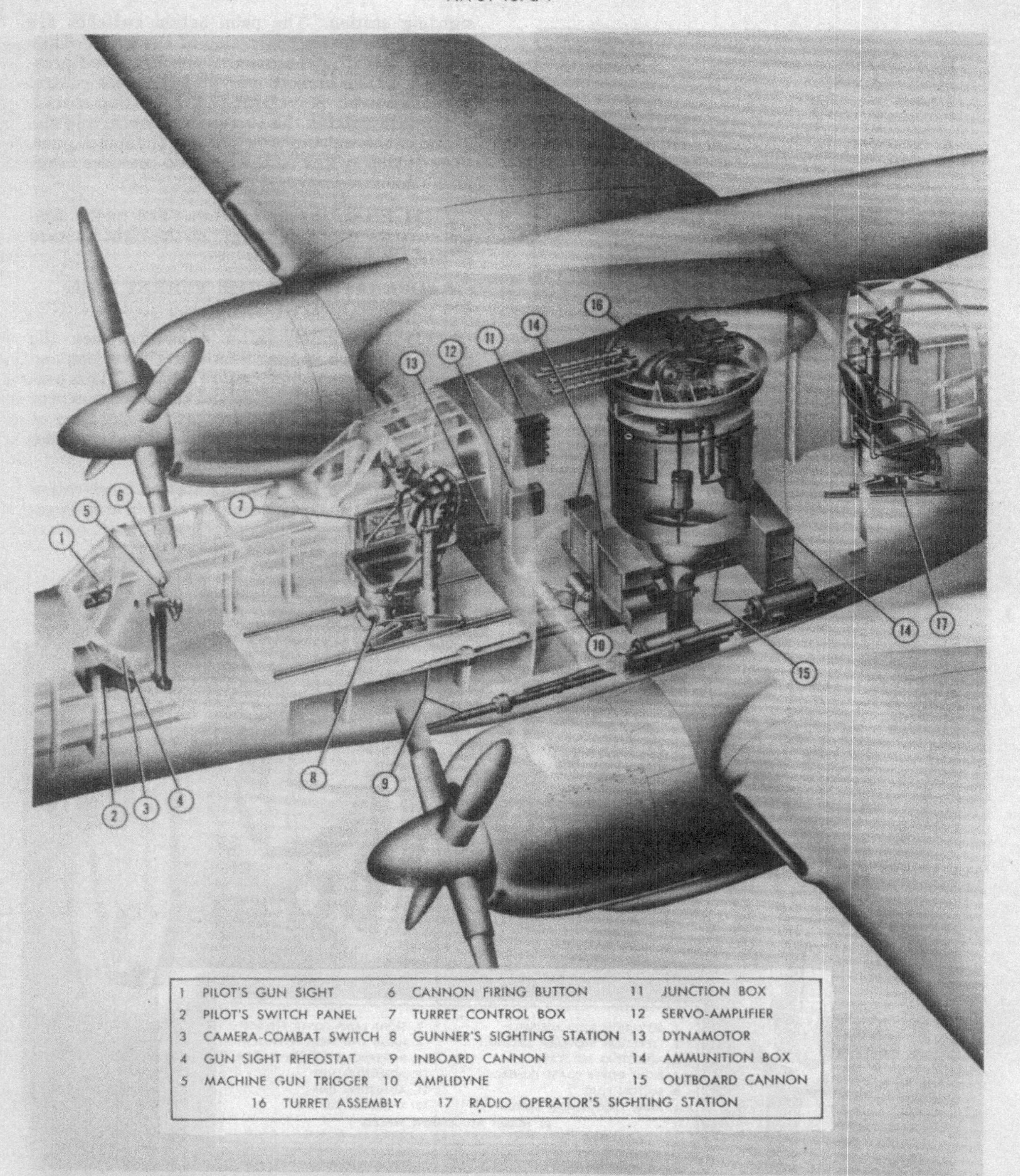


***Figure 48 — Armament Equipment***

ing the sighting arm to the desired position. Unstow the sighting station in azimuth by pulling out the azimuth stowage knob and turning it to fix the lock in the unstowed position. When the operation of the turret and sighting stations is required and, in any case, not until 20 seconds after the operation switch has been placed in the "STANDBY" position, move the operation switch to the "COMBAT" position.

**CAUTION**

The "GUNS" switch should always be returned to the "SAFE" position when the system is not in use.

(3) Depress the palm action switch on the sighting station grip to control the turret. The gunner's sighting station has primary control. The radio operator can control the turret only when the gunner's action switches are open.

(4) To fire the turret guns, depress the trigger switch.

(5) If the turret is to be controlled by the pilot, the gunner and radio operator must open all action switches. The turret will automatically return to the strafing position. Leave the operation switch at "COMBAT" and the "GUNS" switch at "FIRE." To fire the turret guns, press the trigger switch on the right side of the control wheel.

*d.* **ADJUSTMENT OF SIGHTING STATION.**

(1) **HEIGHT ADJUSTMENT.**—To adjust the sight to the required eye level of the individual gunner, loosen the locking wheel near the top of the support column and turn the adjusting wheel until the desired position is obtained. Tighten the locking wheel. (See 7, figure 64.)

(2) **DISTANCE ADJUSTMENT.**—To adjust the sight to the proper distance from the eye of the gunner, loosen the set screw (see 3, figure 47) and adjusting nut (see 10, figure 47) on the flange and tube assembly of the elevation arm. This makes it possible to move the arm in and out. When the desired adjustment is obtained, tighten the set screw and adjusting nut.

*e.* **MOVING THE SIGHTING STATION.**

(1) To move the sighting station fore and aft, release the locking bolts by lifting up the locking bolt handle (see 16, figure 64) at the front of the base. The sighting station now will roll on the track within the limits of the track bumpers. The gunner's sighting station locks in the firing position only. The radio operator's sighting station can be locked in either the firing or radio operating positions.

(2) To rotate the upper portion of the sighting station, pull out the stowage locking knob (see 17, figure 64) and turn it 180 degrees. The station may now be rotated through 350 degrees.

(3) To rotate the seat only, lift up the lever on the left side of the seat and place it in the latched position. The seat is now free to rotate through 360 degrees. To lock the seat, release the lever to the unlatched position and move the seat until the lock engages at either 0 degrees or 180 degrees.

*f.* **STOWING THE SIGHTING STATION.**—Stow before landing to protect the gunner from injury and landing decelerations.

**CAUTION**

Place the "GUNS" switch in the "SAFE" position and the operation switch in the "OFF" position before stowing the sighting station.

(1) **ELEVATION UNIT.**—To stow the elevation unit, lift the elevation arm until the stowage pin (see 2, figure 47) passes the stop (see 1, figure 47) and snaps into the locking clip (see 12, figure 47).

(2) **AZIMUTH UNIT.**—To stow the azimuth unit, rotate the housing to approximately zero degrees azimuth position and turn the stowage locking knobs 180 degrees to the released position. If the locking pin does not snap into the keeper, rotate the housing slowly back and forth until it locks.

## 4. COMMUNICATIONS AND RADIO EQUIPMENT. (See figure 60.)

*a.* **GENERAL.**—The radio equipment installed in the airplane consists of the following sets: AN/ARC-3 command set, SCR-695 identification set, RC-36 interphone, AN/APN-1 altimeter or ACR-718 altimeter, SCR-720 radar, SCR-729 radar, AN/APS-13 radar, AN/ARN-7 radio compass, T-30 microphones, and HS-33 headsets. Three BC-366 jack boxes (see 2, figure 49), one located to the right of the pilot, and one to the right of the gunner, and one to the right of the radio operator, are installed.

(1) The following selector positions are marked on the jack boxes:

*(a)* **"COMMAND."**—This position controls radio set AN/ARC-3 for transmission and reception with the ground or other aircraft.

*(b)* **"COMP."**—In the compass position, radio set AN/ARN-7 is used for listening to ground stations for navigational purposes.

*(c)* "CALL."—This position is a spring-loaded momentary selection for connecting the pilot and crew members, regardless of the switch position on the other jack boxes.

*(d)* "INTER."—The interphone position is discussed in paragraph *b*, following.

*b.* INTERPHONE SYSTEM.

(1) A *standard* interphone arrangement is available to the pilot and crew members as follows:

*(a)* To *listen:* Set all jack box selectors in the "INTER" position. The pilot or crew members can now hear anyone who is talking over the interphone system.

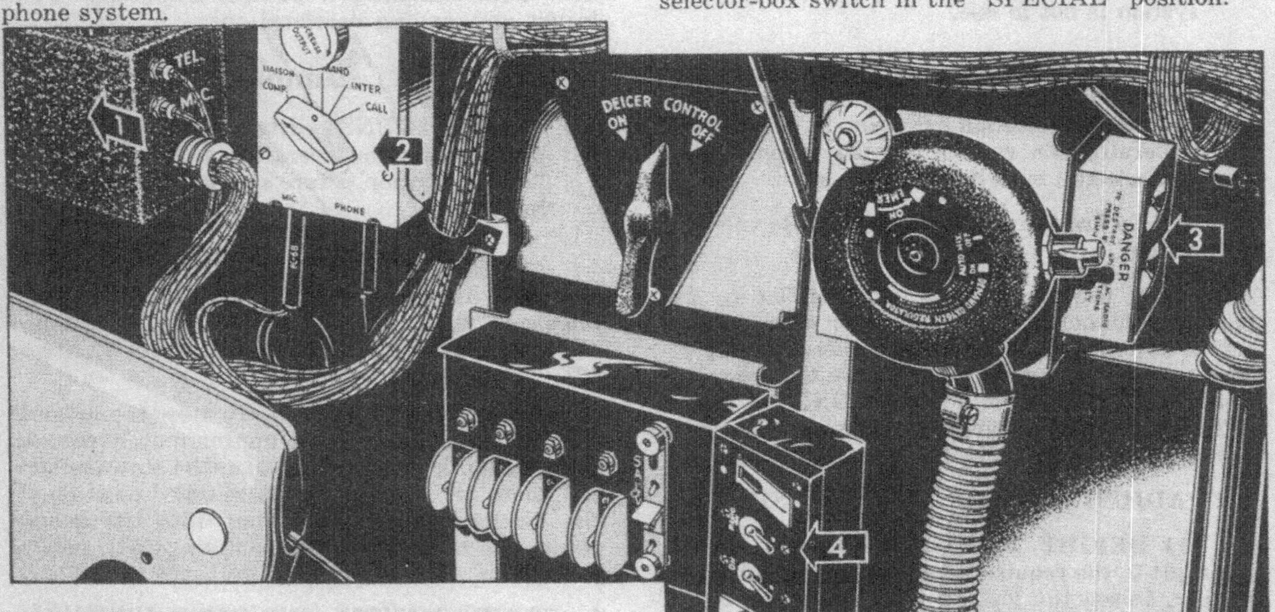


***Figure 49 — Pilot's Radio Controls***

1. AN/ARC-3 BOX
2. BC-366 JACK BOX
3. DESTRUCTOR
4. SCR-695 BOX

*(b)* To *talk* (jack-box selector still in the "INTER" position):

*1.* The pilot actuates the "PRESS-TO-TALK" thumb button on the left side of the control wheel. (See 4, figure 6.)

*2.* The gunner actuates the push-button on the left grip of the sighting station. (See 9, figure 47.)

*3.* The radio operator places his special selector-box switch, on the right side of the compartment, in the "NORMAL" position. (See 16, figure 61.)

a. When acting as rear gunner, the radio operator actuates the push-button on the left grip of the sighting station. (See 9, figure 47.)

b. When acting as radio operator, the radio operator actuates the "PRESS-TO-TALK" foot switch. (See 31, figure 61.)

(2) A *special* interphone arrangement is available as follows:

*(a)* To *listen:* The pilot, gunner or radio operator, can use any selector position on the jack box ("COMMAND," "COMP," "INTER," or "CALL").

*(b)* To *talk:*

*1.* The pilot actuates either the "INTERPHONE" trigger-switch on the left rear side of the control wheel, or the throttle knob interphone button.

*2.* The radio operator places his special selector-box switch in the "SPECIAL" position.

a. When acting as rear gunner, the radio operator actuates the push-button on the left grip of the sighting station.

b. When acting as radio operator, the radio operator actuates the "PRESS-TO-TALK" foot switch.

*3.* The front gunner does not have special facilities under this particular arrangement.

(3) The advantage of the special interphone arrangement is that it provides immediate intercommunication without the loss of time necessary to select identical jack-box positions.

(4) The disadvantage of the special interphone arrangement is that it eliminates "COMMAND" and "COMP" signals when the pilot or radio operator actuate the switches to talk.

(5) It is recommended that the standard interphone arrangement, outlined in paragraph 4 *b* (1), preceding, be used whenever conversation between two parties only is desired.

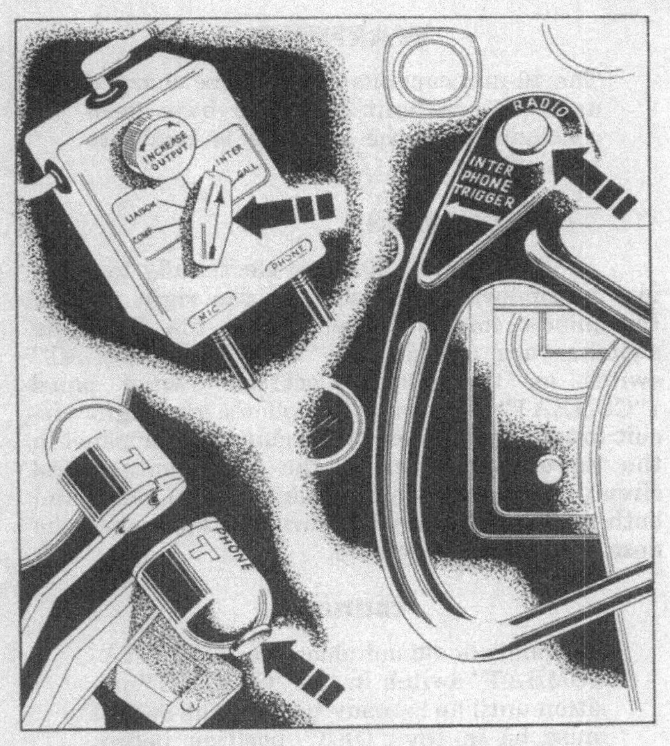


*Figure 50 — Pilot's Interphone Controls*

(6) The "CALL" jack-box position, described in paragraph 4 *a* (1) *(c)*, preceding, can be used by the pilot or crew members instead of the standard or special interphone arrangements, overriding, but not eliminating, "COMP" or "COMMAND" signals, regardless of other jack-box position settings.

*(a)* The disadvantage of the "CALL" setting is that it is spring-loaded, and must be held in position when it is used. Also, the speaker's voice will be mixed together with other conversations on "COMMAND" or reception on "COMP."

*(b)* The advantage of this setting is that intercommunication is obtained without disconnecting a station or stations with which command or compass contact is being maintained.

*c.* **INTERCOMMUNICATION IN CASE OF INTERPHONE AMPLIFIER FAILURE.** — The following alternative is available in case the "INTER" and "CALL" positions are inoperative.

(1) To *listen:* All jack-box selectors must be set on the "COMMAND" position.

(2) To *talk* (jack-box selector still in "COMMAND" position):

*(a)* The pilot actuates the "PRESS-TO-TALK" button on the left side of the control wheel.

*(b)* The front gunner actuates the push-button on the left grip of his sighting station.

*(c)* The radio operator places his special selector-box switch in the "NORMAL" position.

*1.* When acting as gunner, the radio operator actuates the push-button on the left grip of his sighting station.

*2.* When acting as radio operator, the radio operator actuates the "PRESS-TO-TALK" foot switch.

(3) The pilot's interphone throttle-button and "INTERPHONE" trigger-switch on the control wheel, and the radio operator's foot switch and sighting station interphone button, with the special selector-box switch in the "SPECIAL" position, are of no use under this emergency condition.

**Note**

It must be remembered that in using this alternative arrangement for pilot and crew intercommunication, the discussions are broadcast, so that nearby enemy aircraft or ground stations can hear pilot and crew conversations.

*d.* **COMMAND SET.**—AN/ARC-3 equipment is installed in the airplane. The normal operating voltage is 28 volts d-c; however, the set will give emergency operation on voltages between 22 and 32 volts d-c.

(1) **OPERATION.**

*(a)* **STARTING THE EQUIPMENT.** — Push any one of the eight red channel-selector buttons, designated "A" to "H" on the control box (see 1, figure 49). This automatically releases the "OFF" button and applies power to the equipment.

*1.* To transmit:

a. The pilot actuates the "PRESS-TO-TALK" thumb button on the left side of the control wheel.

b. The gunner actuates the push-button on the left grip of the sighting station.

c. The radio operator places his special selector-box switch, on the right-hand side of the compartment, in the "NORMAL" position.

*(1)* When acting as rear gunner, the radio operator actuates the push-button on the left grip of the sighting station.

*(2)* When acting as radio operator, the radio operator actuates the "PRESS-TO-TALK" foot switch.

*(b)* **STOPPING THE EQUIPMENT.**— Push the black "OFF" button and the small metal lock button at the same time. Both of these but-

tons are on the control box. The lock button is provided to prevent the equipment from being turned off accidentally.

*e.* IDENTIFICATION EQUIPMENT.—Identification of the airplane in flight is provided for by the installation of SCR-695 equipment. The equipment consists of a radio receiver, RC-255A radio control equipment, and a destructor unit.

(1) OPERATION.

*(a)* To start the equipment, throw the ON-OFF switch on the RC-255-A control box or the pilot's remote control box (see 4, figure 49) to "ON."

*f.* MICROPHONES. — Each crew member is provided with a T-30 variable-sensitive, throat microphone and an HS-33 headset.

**5. PILOT'S COMPARTMENT.**

*a.* GUNNERY EQUIPMENT.

(1) FIXED CANNON.

*(a)* DESCRIPTION.—The airplane is equipped with four 20-mm cannons, mounted on supports beneath the crew-nacelle floor structure. Each cannon is provided with a 200-round ammunition box. The cannons are charged manually on the ground. If one or more of the cannons should jam, the others will continue to operate.

**WARNING**

The 20-mm cannons must not be charged until all pre-flight activities have been concluded and the airplane is ready to take off.

*(b)* CONTROLS.

*1.* Fire is controlled electrically by a firing button-switch, installed in the right side of the pilot's control wheel, when the following switches are closed: the "CAMERA-COMBAT" switch on the pilot's electrical switch panel ("COMBAT" position), the pilot's gun sight circuit breaker switch on the circuit breaker panel in the turret compartment, the armament control circuit breaker on the generator control panel, and both micro switches which are actuated when the nose gear doors are closed.

**CAUTION**

The pilot should not place the "CAMERA-COMBAT" switch in the "COMBAT" position until he is ready to fire. The switch must be in the "OFF" position before landing the airplane.

(2) GUN SIGHT.

*(a)* DESCRIPTION.—The pilot is provided

*Figure 51 — Night Binocular*

with a Type LY-3N gun sight mounted above the instrument panel. (See 2, figure 6.)

*(b)* **OPERATION.**—The pilot's gun-sight lamp is operated by a switch located on the right side of the gun sight, and controlled by a rheostat mounted on the pilot's electrical panel.

(3) **NIGHT BINOCULAR.** (See figure 51.)

*(a)* **DESCRIPTION.**—The night binocular is a combination of 5.8 power night glasses and an optical gunsight. It is mounted on an arm which permits it to swing into operating position aft of the pilot's armor glass. Normally it is stowed at the aft end of the track on the compartment left rail.

*(b)* **TO BRING THE BINOCULAR TO OPERATING POSITION.**

*1.* Reach back with the left hand and grasp the vertical handle located at the lower forward corner of the binocular carriage.

*2.* Pull forward on the handle until the carriage latches at the forward end of the track.

*3.* With the right hand, release the binocular from the carriage by pushing the release at the upper rear corner of the carriage.

*4.* Swing the binocular around until the cone plug almost touches the cone socket.

*5.* Push in on the handle located at the top center of the binocular support casting and seat the cone plug in the socket.

*6.* Lock the binocular in place by a 90-degree rotation of the handle.

*(c)* **TO OPERATE THE BINOCULAR.**

*1.* Grasp the operating handle with the left hand.

*2.* Depress the small lever attached to the main handle. This allows the binocular to be swung in azimuth and elevation.

*3.* Release the handle in any position to lock the binocular in that position.

**Note**

The binocular may be rotated 30 degrees each side of the center line, 5 degrees below the horizontal, and 25 degrees above the horizontal.

*4.* Regulate the brightness of the reticle by means of the rheostat located in the binocular.

*(d)* **TO STOW THE BINOCULAR.**

*1.* Unlock the binocular in the operating position by rotating the handle 90 degrees to the right.

*2.* Pull on the handle located at the top center of the binocular support casting to unseat the cone plug.

*3.* Stow the binocular in the carriage.

*4.* Slide the binocular back until it latches at the aft end of the track.

(4) **COMBAT CAMERA.** — An electrically-operated camera which works in conjunction with the cannons, is installed in the crew-nacelle nose. The camera operates for the length of time the pilot's cannon trigger is depressed, plus the time for which the overrun control is set.

*(a)* **OPERATION.**

*1.* Place the switch on the pilot's electrical switch panel in the "COMBAT" position if the camera is to be used in conjunction with the cannon, or in the "CAMERA" position if it is to be used alone.

*2.* Depress the 20-mm cannon-firing switch on the control wheel.

*3.* Return the switch on the pilot's electrical switch panel to "OFF" when the operation is completed.

(5) **PILOT'S TURRET CONTROL.** — The pilot's turret control consists solely of a firing trigger installed in the right side of the control wheel. The pilot can operate the turret only when it is in the strafing position, and both gunners have relinquished control.

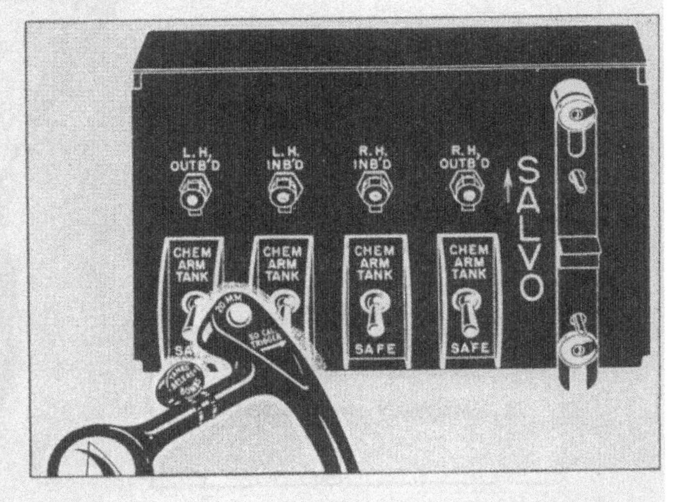


*Figure 52 — Bomb and Drop Tank Controls*

*b.* **BOMBING AND CHEMICAL TANK EQUIPMENT.**—Provisions are made for the installation of external droppable chemical tanks, or bombs, under each inner and each outer wing panel.

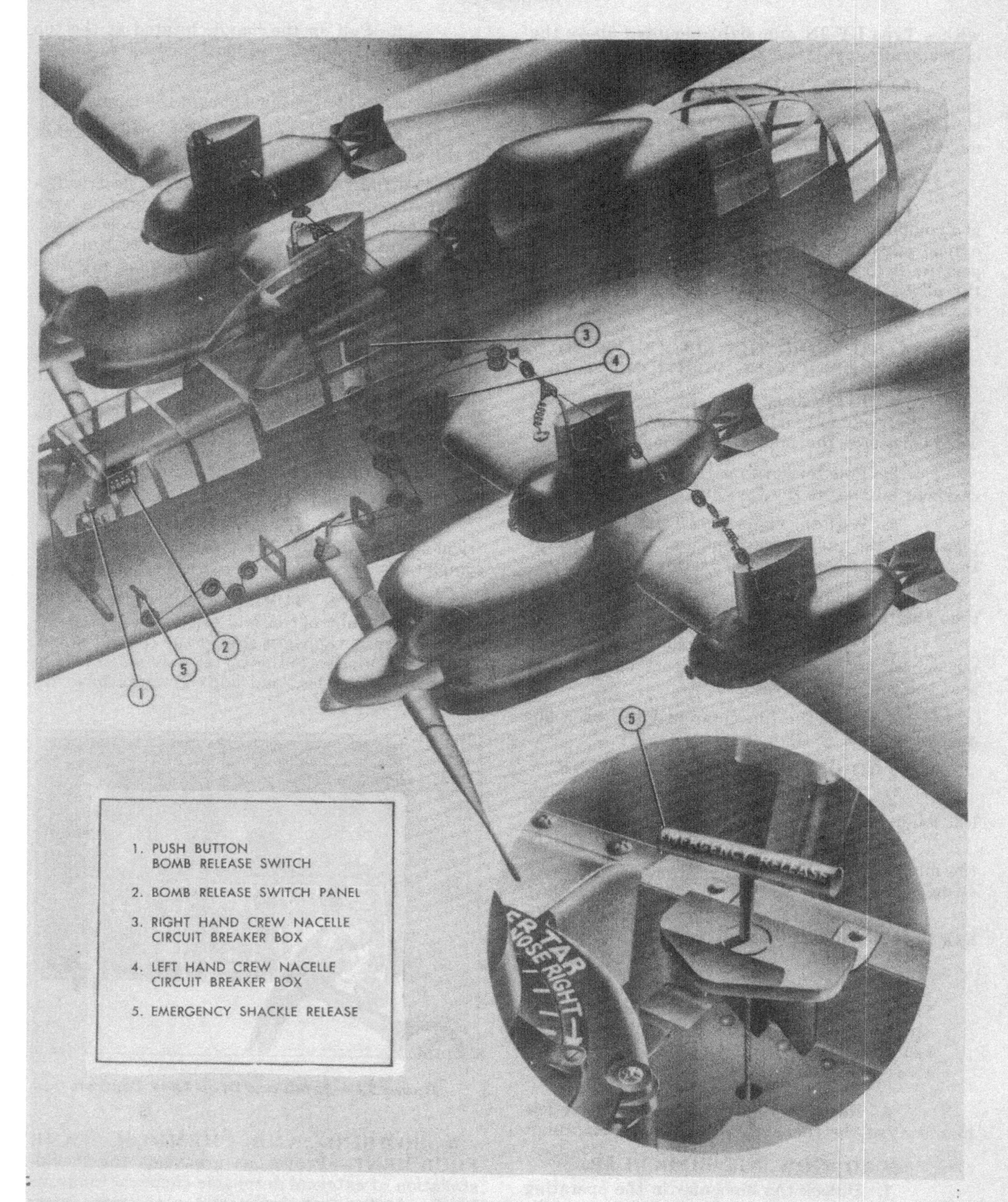


***Figure 53 — Bomb and Drop Tank Control System***

(1) CONTROLS. (See figure 52.)—The bomb selector switch panel is mounted on the right rail, opposite the pilot. The panel contains six switches and four indicator lights. Each shackle is controlled by an on position marked "CHEM-ARM-TANK." Each of these switches is protected against accidental release by a guard. An indicator light, mounted above the control switch for each shackle, comes on when the switch is in the "ON" position and the unit has not been released from the shackle. The other two switches, marked "SALVO," are linked together and controlled by a single handle. A push-button type switch, labeled "TANK-RELEASE-BOMBS," is mounted on the upper right spoke of the control wheel and provides the actual release control. A switch-type circuit breaker, marked "BOMB CONTROL," is located on the generator control panel.

(2) OPERATION.

*(a)* TO RELEASE BOMBS ARMED.

*1.* On the generator control panel, switch "ON" the circuit breaker marked "BOMB CONTROL."

*2.* On the bomb selector switch panel, turn the switch corresponding to the desired bomb to "ARM." Immediately, as the selector switch is displaced from "SAFE," the indicator light for the selected station will go on.

*3.* Press the release button on the control wheel. As soon as the bomb has left the airplane, the indicator light will go out, indicating that the bomb has cleared the shackle.

*(b)* TO RELEASE BOMBS SAFE.—Follow the procedure in paragraph *(a)*, preceding, except in step *2*, leave the selector switch in the "SAFE" position.

*(c)* TO RELEASE CHEMICALS FROM CHEMICAL TANKS.

*1.* On the generator control panel, switch "ON" the circuit breaker marked "BOMB CONTROL."

*2.* On the bomb selector switch panel, turn the switch corresponding to the desired tank to "CHEM." Immediately, as the switch is displaced from "SAFE," the chemicals will be released from the tank and the indicator light for the selected station will go on. The indicator light will remain on until either the selector switch is returned to "SAFE" or the release button on the control wheel is pressed.

*(d)* TO RELEASE UNITS BY SALVO.

*1.* On the generator control panel, switch "ON" the circuit breaker marked "BOMB CONTROL."

*2.* On the bomb selector switch panel, turn the handle labeled "SALVO" up. Immediately, as the switches are displaced, the indicator lights for all stations will go on.

*3.* Press the button on the control wheel. As soon as each shackle opens, the corresponding indicator light will go out.

*c.* DE-ICING AND ANTI-ICING EQUIPMENT.

(1) PROPELLER ANTI-ICER.

*(a)* DESCRIPTION. — Each propeller is equipped with a spinner and a slinger ring. A 6 US (5 Imperial) gallon anti-icer fluid tank with electrically-driven pump is located beneath the crew-nacelle floor in the radio operator's compartment.

*(b)* CONTROLS.—A push-button type circuit breaker, labeled "ANTI-ICER," is located on the generator control panel. The rheostat control switch and anti-icer gage are located forward of the control column, on the floor.

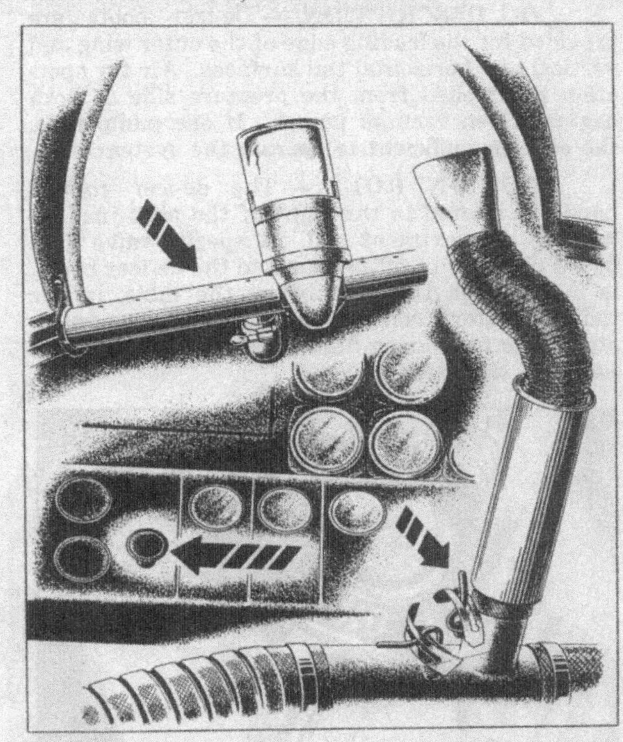

*Figure 54 — Pilot's Windshield Defroster*

(2) WINDSHIELD DEFROSTERS. (See figure 54.)

*(a)* DESCRIPTION. — A fixed defroster tube is provided for the inside of the pilot's windshield. A flexible tube, installed in a container to the right of the pilot's seat, provides additional means of defrosting the pilot's enclosure.

(*b*) **CONTROLS.**—A "CABIN AIR" control and a "DEFROST" control are located at the base of the pilot's flexible defrost tube. The "PULL TO DEFROST WINDSHIELD" control is on the auxiliary control panel.

(*c*) **OPERATION.**

*1.* **FIXED DEFROSTER.** — Place the "CABIN AIR" control lever on the gunner's cabin air outlet in the "OFF" position. Place the "CABIN AIR" and "DEFROST" control levers at the base of the pilot's flexible defrost outlet in the "OFF" position. Pull out the "WINDSHIELD DEFROST" control on the auxiliary control panel.

*2.* **FLEXIBLE DEFROST.** — Place the "CABIN AIR" controls on the gunner's cabin air outlet and at the base of the pilot's fixed defroster outlet in the "OFF" position. Place the "DEFROSTER" lever at the base of the pilot's flexible defrost outlet in the "ON" position.

(3) **DE-ICER BOOTS.**

(*a*) **DESCRIPTION.** — De-icer boots are provided for the leading edge of the outer wing and vertical and horizontal tail surfaces. Air for operation is supplied from the pressure side of both engine-driven vacuum pumps. If one pump fails, the other is sufficient to operate the system.

(*b*) **CONTROLS.** — The de-icer control handle is located to the right of the pilot just below the compartment rail. A special valve (see figure 55) to shut off pressure to the de-icer boots, is installed on the floor below the table, in the radio operator's compartment. If the vacuum instruments fail during flight, break the safety wire and turn the valve to "OFF."

*Figure 55 — De-Icer Shut-Off Valve*

*d.* **RADIO EQUIPMENT.**—In addition to the radio controls discussed in paragraph 4, preceding, the pilot operates a radio compass, an absolute altimeter, and the AN/APS-13 equipment.

(1) **RADIO COMPASS.** — The AN/ARN-7 radio compass equipment includes a radio control box and a pilot's indicator. The control box is located on the compartment right rail opposite the pilot, and the indicator is mounted on the compartment rail forward of the controls.

**Note**

Operate the AN/ARN-7 compass in flight only. The radio operator's and pilot's SCR-720 must be in "OFF" or in the extreme counterclockwise position unless the radar is being used.

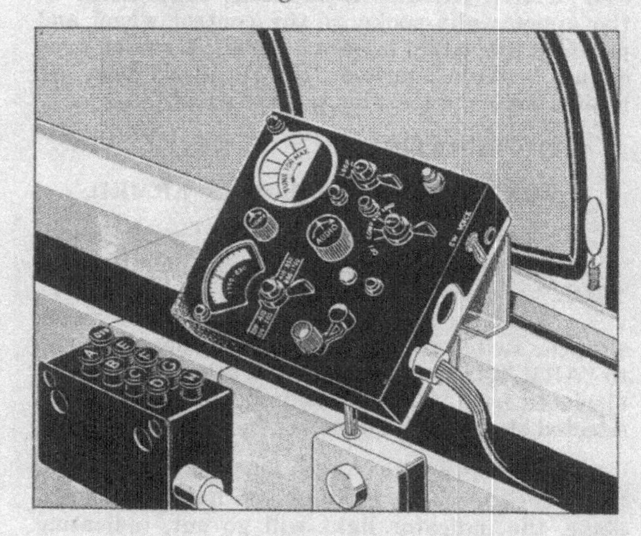

*Figure 56 — AN/ARN-7 Controls*

(2) **ABSOLUTE ALTIMETER.**—The control box for the AN/APN-1 equipment is mounted above the automatic pilot pressure switch to the right of the pilot. The indicator is located on the top, right-hand side of the main instrument panel.

(*a*) **CONTROLS.** — The control box contains two scales; one from 0 to 400 feet and one from 0 to 4000 feet. Adjacent to the scales are three indicator lights. The red light shows when the airplane is at an altitude below the scale setting, the green when it is above the setting, and

*Figure 57 — Absolute Altimeter*

the white when it is directly on the setting. A rheostat, with "BRIGHT" and "DIM" positions, controls the intensity of the indicator lights. A limit switch is mounted on the control box, and a knob-type power switch is mounted on the main instrument panel by the indicator.

*(b)* **OPERATION.**—Turn the power switch "ON." Allow the units to heat for one minute.

**Note**

When the aircraft is resting on the ground, the indicator may not indicate zero altitude. Do not adjust the equipment to obtain a zero reading.

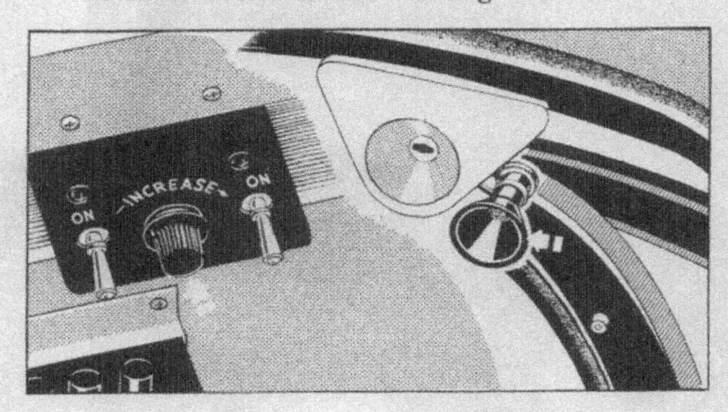


*Figure 58 — AN/APS-13 Equipment*

(3) **AN/APS-13 EQUIPMENT.**—The equipment provides a means of warning against the approach of aircraft from the rear. It includes a control panel, a warning bell, and an indicator light. The control panel is installed below the right rail, opposite the control column. The warning bell is mounted below the anti-icer instrument panel. The indicator light is attached by a bracket to the side of the night binocular forward-position lock.

*(a)* **CONTROLS.**—The controls consist of a "TEST" switch and an operating switch, both having "ON" and "OFF" positions, and a rheostat to control the intensity of the indicator light.

*(b)* **OPERATION.**—To turn on the equipment, place the operating switch in the "ON" position. Allow the equipment to warm up for three minutes.

*(c)* **TEST.**—Hold the test switch up in the "ON" position against the spring tension. If the red indicator light illuminates and the warning bell rings, the equipment is operating properly.

*e.* **ADJUSTMENT OF PILOT'S SEAT.**

(1) The upper aft control lever, located on the right side of the pilot's seat, allows the seat to be adjusted vertically. The lower forward control lever, at the right side of the pilot's seat, permits the seat pan to be tilted up or down.

(2) The complete seat can be tilted back to facilitate entrance and exit. The seat is released for tilting by pressing down the knob located on the back of the seat at the top and to the right of the center.

**6. RADIO OPERATOR'S COMPARTMENT.**
(See figure 61.)

*a.* **COMMUNICATIONS.** — In addition to the radio equipment discussed in paragraph 4, preceding, the radio operator operates the SCR-720 and SCR-729 radar equipment and the SCR-718 altimeter.

(1) **SCR-720 RADAR EQUIPMENT.** — The equipment includes a control box, a pilot's indicator, a radio operator's indicator, and a synchronizer.

*(a)* **TO START EQUIPMENT.**—Push the "POWER" switch "ON" button.

*(b)* **TO STOP EQUIPMENT.** — Push the "POWER" switch "OFF" button.

(2) **SCR-729 RADAR EQUIPMENT.** — The equipment includes a receiver-transmitter, an indicator, and a radio control box.

*(a)* **TO START THE EQUIPMENT.**

*1.* Place the "D1-1R" switch in the mid-position.

*2.* Place the "ON-OFF" switches in the "ON" position.

*3.* Allow three minutes for the equipment to warm up.

*(b)* **TO STOP THE EQUIPMENT.**—Turn the "ON"-"OFF" switches "OFF."

(3) **SCR-718 ALTIMETER.** — The SCR-718 is an alternate installation to the AN/ARN-1 altimeter. Both sets are not installed in the airplane at the same time.

*(a)* **OPERATION.** — Start the equipment by turning the "REC. GAIN" control knob clockwise one-half turn. Allow the set to warm up for five minutes. Adjust the "CIRCLE SIZE" control so that the circle on the indicator is entirely visible.

*b.* **FLEXIBLE DEFROSTER.**—A flexible defroster tube is installed on the left side of the radio operator's compartment. To operate, place the "CABIN AIR" control lever in the "OFF" position and the "DEFROST" control "ON." Remove the tube from the clip and container.

*c.* **SUIT HEAT CONTROLS.**—A suit heat control box is installed under the radio operator's table

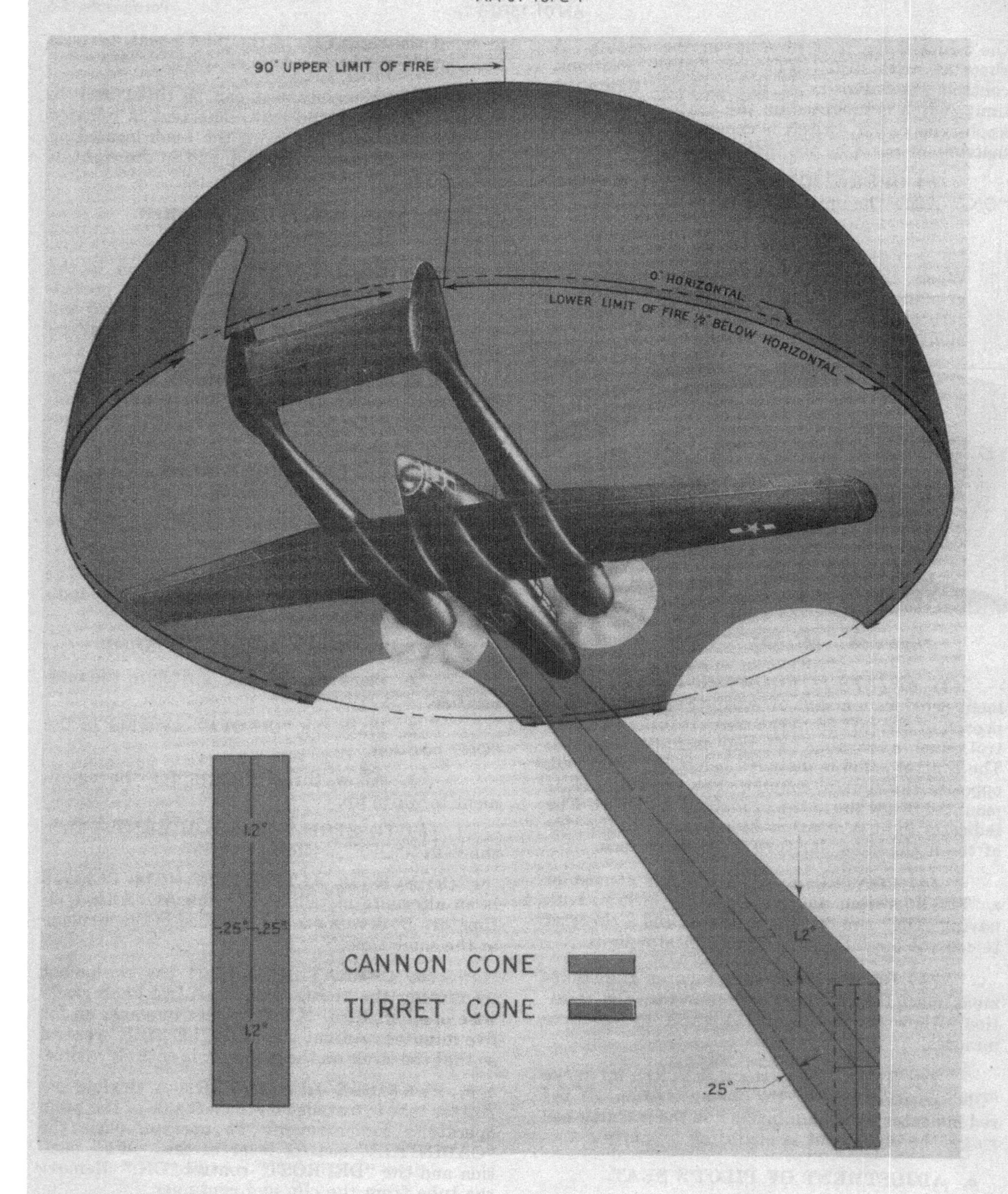


*Figure 59 — Cones of Fire*

1. RADIO ALTIMETER
2. AN/APS-13 INDICATOR LIGHT
3. COMPASS INDICATOR
4. AN/APN-1 CONTROLS
5. RADIO COMPASS CONTROLS
6. AN/APS-13 CONTROLS
7. AN/ARC-3 CONTROLS
8. INTERPHONE JACK BOX
9. SCR-695 CONTROLS
10. SCR-695 DETONATOR
11. AN/APS-13 WARNING BELL
12. RIGHT-HAND CIRCUIT BREAKER BOX
13. SCR-720 CONTROL BOX
14. SCR-720 JUNCTION BOX
15. SCR-720 INDICATOR
16. SCR-718 INDICATOR
17. SCR-729 RECEIVER TRANSMITTER
18. SCR-729 INDICATOR
19. SCR-729 CONTROL BOX
20. SCR-720 RECTIFIER
21. SCR-720 VOLTMETER
22. SCR-720 SYNCHRONIZER
23. SCR-695 CONTROL BOX (SELECTOR)
24. SCR-720 INVERTER
25. SCR-695 CONTROL BOX (POWER)
26. ELECTRICAL JUNCTION BOX
27. SCR-718 OR AN/APN-1
28. PUSH TO TALK SWITCH
29. INTER PHONE DYNAMOTOR
30. RADIO COMPASS
31. RADIO JUNCTION BOX
32. AN/ARC-3 EQUIPMENT
33. SCR-695 TRANSMITTER-RECEIVER
34. AN/APS-13 TRANSMITTER-RECEIVER

***Figure 60 — Radio and Communication Equipment***

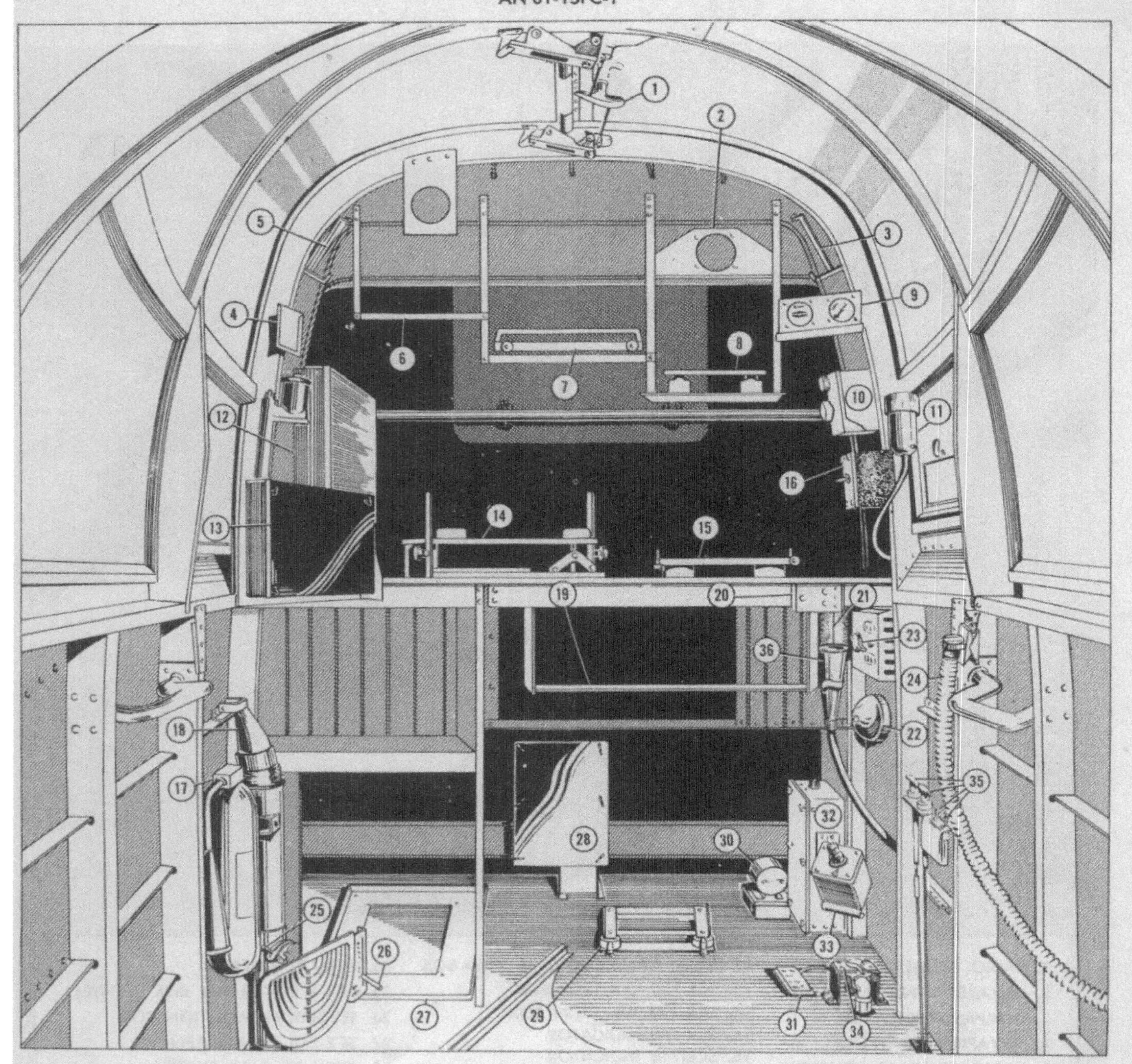

***Figure 61 — Radio Operator's Compartment***

1. CLAM SHELL RELEASE
2. VOLTMETER
3. SCR-695 CONTROL BOX MOUNT
4. CARD HOLDER
5. SCR-729 CONTROL BOX MOUNT
6. SCR-718 INDICATOR MOUNT
7. SCR-729 INDICATOR MOUNT
8. SCR-720 RECTIFIER MOUNT
9. RADIO OPERATOR'S OXYGEN INST.
10. RADIO OPERATOR'S BC-366 JACK BOX
11. SPOT LIGHT
12. JUNCTION BOX (SCR-720)
13. SCR-720 CONTROL BOX
14. SCR-720 INDICATOR MOUNT
15. SCR-720 SYNCHRONIZER MOUNT
16. SPECIAL SELECTOR BOX
17. FIRE EXTINGUISHER
18. FLEXIBLE DEFROST TUBE
19. SCR-718 OR AN/APN-1 MOUNT
20. SPARE LAMPS
21. INTERPHONE AMPLIFIER
22. ALARM BELL
24. SUIT HEATER
24. OXYGEN TUBE
25. DEFROST CONTROL
26. "CABIN AIR" CONTROL
27. RADIO COMPASS MOUNT (AN/ARN-7)
28. JUNCTION BOX (RADIO)
29. SCR-729 REC.-TRANS. MOUNT
30. DYNAMOTOR (INTERPHONE)
31. "PUSH-TO-TALK" FOOT SWITCH
32. JUNCTION BOX (ELECTRICAL)
33. EXTENSION LIGHT
34. DE-ICER SHUT-OFF VALVE
35. AFT HEATER CONTROLS
36. RELIEF TUBE

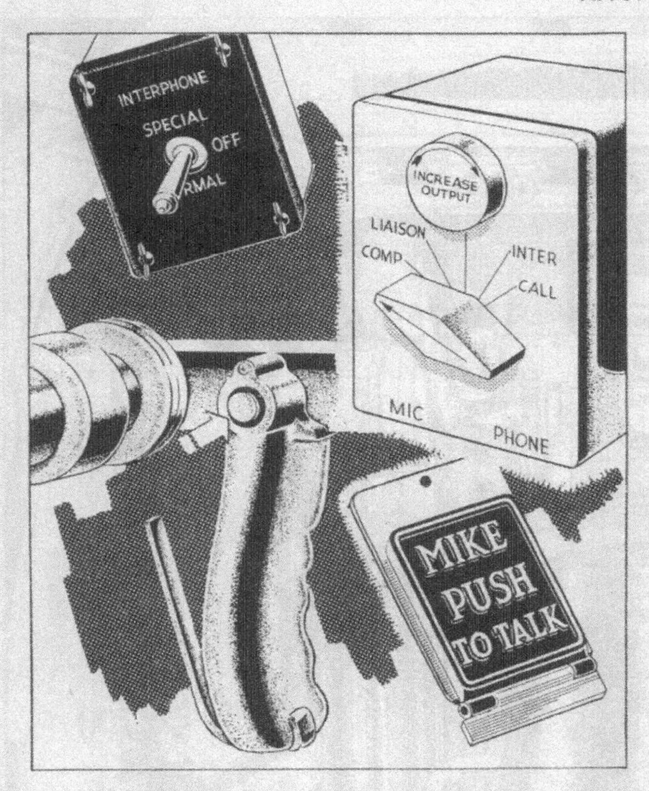


**Figure 62 — Radio Operator's Inter Phone Controls**

on the left side of the compartment. It contains a plug receptacle and a rheostat control for adjusting current output.

*d.* **VENTILATOR.** — A ventilating window is installed in the radio operator's enclosure.

## 7. GUNNER'S COMPARTMENT.

(See figures 63 and 64.)

*a.* **WINDSHIELD DEFROSTER.**—A fixed defrost tube is installed along the base of the gunner's windshield and a flexible defrost tube is in a container to the right of the gunner's seat.

(1) **CONTROLS.**—The "CABIN AIR" control is mounted on the cabin air outlet, and the "DEFROST" control at the base of the gunner's flexible defrost tube.

(2) **OPERATION.** — To operate the fixed windshield defroster, attach the flexible defroster tube to fixed-defroster adapter. Close the "CABIN AIR" control and turn the "DEFROST" control "ON." To operate the flexible defroster, remove the tube from the clip and container, close the "CABIN AIR" control and turn the "DEFROST" control "ON."

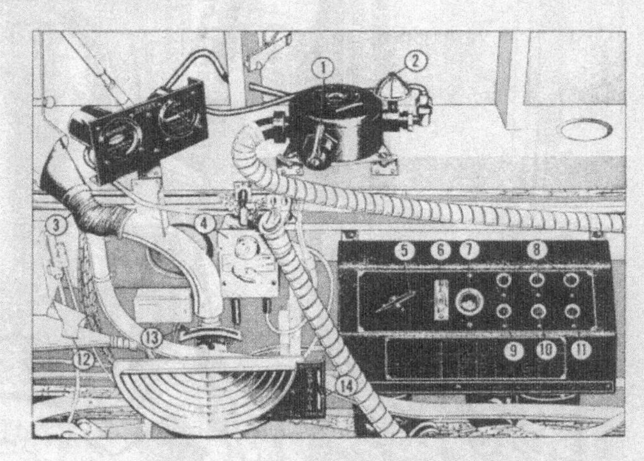

**Figure 63 — Gunner's Compartment — Right**

1. OXYGEN AUTO-MIX CONTROL
2. OXYGEN EMERGENCY VALVE CONTROL
3. GUNNER'S FLEXIBLE DEFROST TUBE
4. GUNNER'S BC-366 JACK BOX
5. TURRET OPERATION SWITCH
6. TURRET "GUNS" SWITCH
7. "TURRET POWER" CIRCUIT BREAKER
8. INDICATOR LIGHTS
9. "POWER CONTROL" CIRCUIT BREAKER
10. "AUXILIARY" (PILOT) CIRCUIT BREAKER
11. "COMPUTOR" CIRCUIT BREAKER (NOT USED)
12. GUNNER'S RELIEF TUBE
13. GUNNER'S DEFROST CONTROL
14. GUNNER'S "CABIN AIR" CONTROL

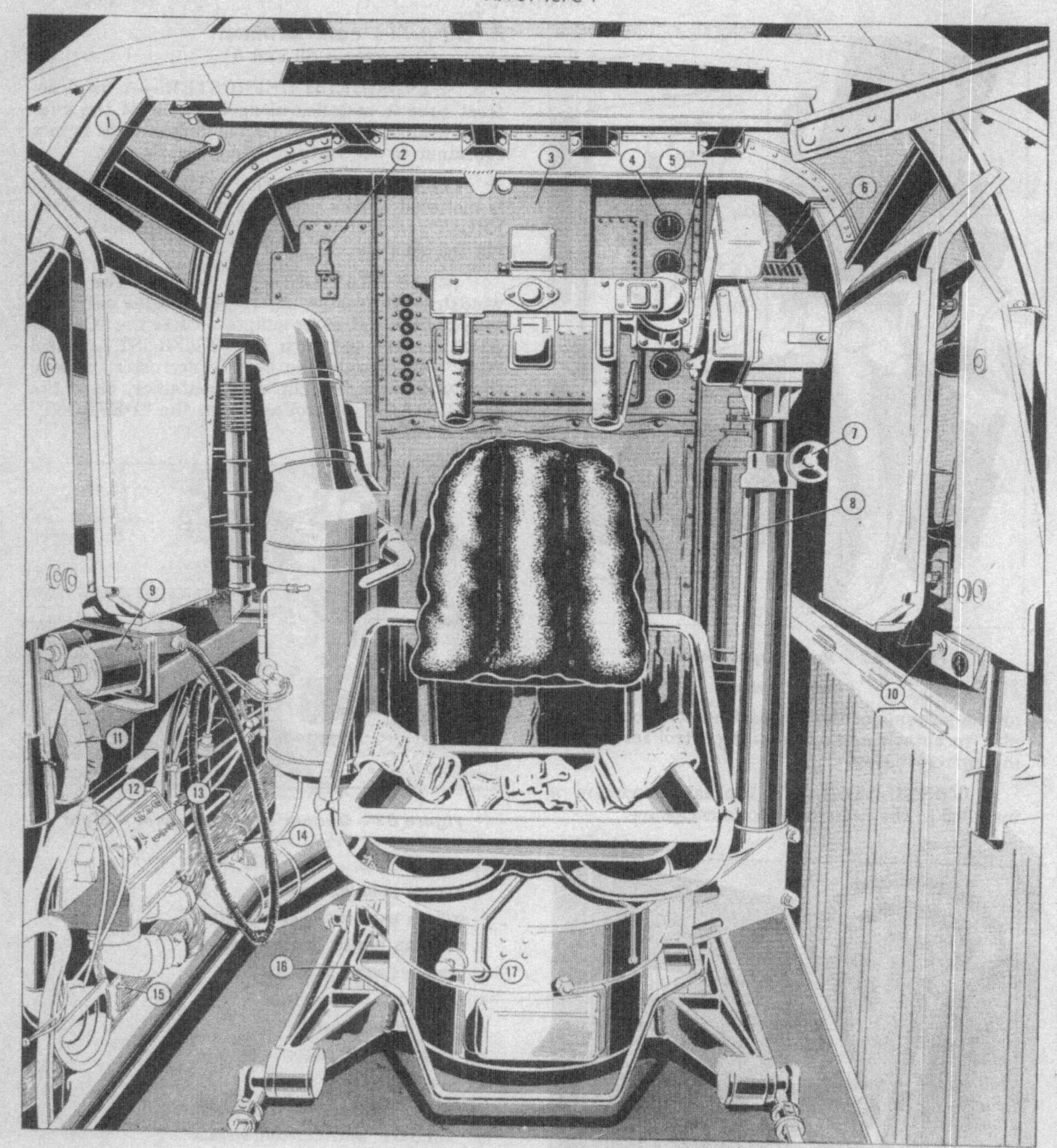


***Figure 64 — Gunner's Compartment***

1. EMERGENCY EXIT RELEASE
2. ADAPTER GOGGLES BRACKET
3. PYROTECHNIC PISTOL AND FLARES
4. RESERVOIR PRESSURE GAGE
5. ACCUMULATOR PRESSURE GAGE
6. SUIT HEATER
7. ELEVATION ARM ADJUSTING WHEELS
8. FIRE EXTINGUISHER
9. GUNNER'S OXYGEN INSTRUMENTS
10. AIR BRAKE GAGE AND FILLER
11. GUNNER'S DEFROST TUBE
12. GUNNER'S BC-366 JACK BOX
13. TURRET CONTROL BOX
14. COLD AIR CONTROL
15. "CABIN AIR" CONTROL
16. LOCKING BOLT HANDLE
17. AZIMUTH STOWAGE LOCK
18. SURGE CHAMBER PRESSURE GAGE

# SECTION VI EXTREME WEATHER OPERATIONS

## 1. ARCTIC OPERATION.

### a. BEFORE ENTERING THE PILOT'S COMPARTMENT.

(1) TIRES AND BRAKES. — Move the airplane enough to determine if the tires are frozen to the ground. If they are, thaw them out. If the airplane moves, but the wheels slide, the brakes are frozen. Thaw them out.

(2) COVERS.—Remove all except the engine nacelle covers unless snow, sleet, or freezing rain is falling, in which case leave all covers on until just before take-off.

(3) PROPELLERS. — Remove any accumulation of snow or ice by hand or a hot air blast heater.

(4) INSTRUMENTS. — If necessary, apply heat to the instruments which operate independently of the engines, to make the normal pre-flight check.

(5) STARTERS. — Use the hand crank to check each starter for freedom of movement.

(6) DRAINS.—In addition to the normal check of fuel drains, check the oil tank sump drains. If no oil flows from the drain cock when it is opened, ice is present and must be thawed out. Check the valves on the Y-drain fittings.

**CAUTION**

Never start the engines in cold weather until these valves have been checked and it is certain that the oil flows freely.

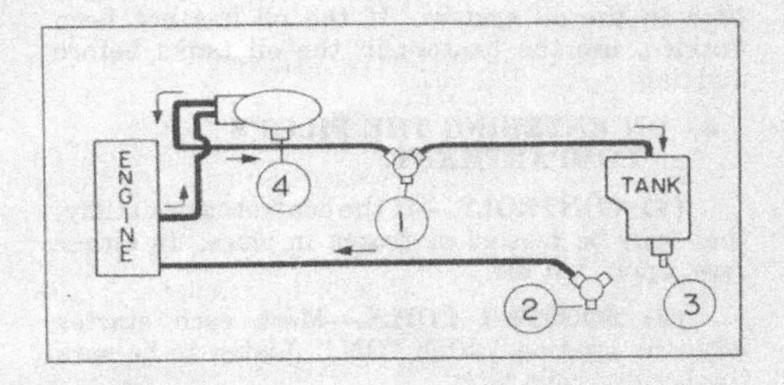

(7) **VENT LINES.**—Check all vent lines to make sure that they are free from ice.

(8) **SNOW AND FROST.**—Look for and remove all accumulations of snow from the airplane either by hand or with compressed air. Remove all frost with a hot air blast before take-off.

(9) **BATTERIES.**—Batteries should be warm before the engine is started. Keep them indoors between flights if possible.

(10) **OIL.**—If oil has been removed from the airplane, heat it to at least 70°C before putting it back in the oil system. If the oil has not been drained, use the heaters in the oil tanks before starting.

*b.* **ON ENTERING THE PILOT'S COMPARTMENT.**

(1) **CONTROLS.**—If the controls work stiffly, they may be frosted or frozen in place. In either case, apply hot air.

(2) **BOOSTER COILS.**—Mesh each starter with the ignition switch "ON." Listen to be sure the booster coils buzz.

(3) **WINDSHIELDS AND WINDOWS.** — If frost has formed on the inside of the glass surfaces, warm the compartment with blast heaters and wipe off the frost as it melts. Once they are clean and dry, open the windows until just before take-off.

*c.* **STARTING ENGINES.**

**WARNING**

Before attempting to start or even to heat the engines, make certain that adequate $CO_2$ fire extinguishers are available.

(1) **PREHEATING ENGINES.**

*(a)* If the outside air temperature is below −18°C (0°F), apply preheat to the engines before attempting to start them. If the engines have sufficient residual heat from the previous flight to indicate any cylinder-head temperature at all, preheating should not be necessary.

*(b)* Pre-heating engines may take as long as three hours, depending on the temperature and equipment available.

*(c)* Use hot air blast heaters, Type D-1 or F-1, if they are available. Be especially careful in using makeshift heaters. Open flames are particularly hazardous because fuel vaporizes from the oil and comes out the engine breathers. Protect all rubber and fabric hoses from temperatures above 121°C (250°F) with asbestos baffles or metal sheets.

(2) **CARBURETOR PRE-HEAT.** — As soon as the engines are warmed externally, unbutton and loosen the engine covers. See that the ignition switch is "OFF," the mixture control in "IDLE CUT-OFF," and the throttle wide open. With the inboard carburetor air intake duct tightly covered, direct a hot air blast into the outboard carburetor air intake duct. While hot air is being directed into the outboard intake duct, turn the engine over by hand through at least six complete revolutions. This heats and dries the spark plugs, the induction system, and the inside of the cylinders. Repeat the process for the other engine. As soon as the preheating of both engines is completed, remove the engine covers.

(3) **PRIMING.**

**Note**

The time required for adequate priming varies from five to more than 30 seconds, depending on the outside air temperature. Underpriming causes more difficulties than overpriming.

*(a)* After the starter is in "MESH," hold the primer switch "ON" until the engine runs. It may be necessary to run the engine on the primer until it is firing on all cylinders. If backfiring occurs, the engine is underprimed. If the engine is overprimed, keep the mixture control in "IDLE CUT-OFF" and advance the throttle to the full "OPEN" position for three seconds. Continue to turn the engine over with the starter.

*d.* **WARM-UP.** — Follow the normal warm-up procedure, except when extremely cold temperatures make it necessary to dilute the oil further in an emergency.

(1) **CARBURETOR AIR HEAT.**—Use carburetor air heat during warm-up when outside temperatures are −20°C (−4°F), or below. Use it also on the ground when conditions are conducive to the formation of carburetor ice. Be sure to turn it off before take-off. If left on, it overheats the engines, causes loss of power, and will probably produce detonation.

(2) **ACCESSORIES.**—Check the operation of the de-icer boots, cowl flaps, and other accessories in the normal manner. Make sure the compartment heaters are turned off prior to take-off.

*e.* **TAKE-OFF.**

(1) If there is danger of propeller icing, turn on the anti-icing control for at least a minute at low engine rpm, just before take-off.

(2) Be sure that carburetor pre-heat and both compartment heaters are "OFF" before starting the take-off.

(3) Always use one-third flaps (20 degrees) in take-offs on snow or ice.

(4) Once the airplane is in the air, turn both compartment heaters on immediately.

*f.* **CRUISING.**

(1) **CARBURETOR ICING.**—When the carburetor air temperature is between −10°C to +10°C in an atmosphere of high humidity or free moisture, carburetor icing is likely to occur. Under such atmospheric conditions, a carburetor air temperature between 10°C and 40°C should be maintained. Use carburetor pre-heat for two minutes every half hour during flight if carburetor icing is suspected.

(2) **LANDING GEAR.**—In flight, operate the landing gear system occasionally to make certain it is not freezing in the "UP" position. Make the first check shortly after take-off.

(3) **TURRET.**—Operate the gun turret frequently, but not continuously, during flight, to prevent sluggishness or freezing.

(4) **DE-ICER SYSTEM.** — If one vacuum pump fails, the entire system will work effectively on the other. If both pumps fail, or a leak in the line prevents the de-icer boots on one wing from inflating, shut off the de-icer system completely.

*g.* **APPROACH AND LANDING.**

(1) **FLAPS.**—When landing on ice- or snow-covered runways, come in with flaps fully extended. This not only insures the slowest possible landing speed but shortens the landing roll and decreases the use of the brakes.

(2) **HEADING.** — Land as nearly into the wind as possible. Cross-wind landings are more hazardous on slippery surfaces.

(3) **ELECTRICAL EQUIPMENT.**—Turn off electric suit heaters at least one minute before the final approach, and make every effort to reduce the use of electrical equipment to save the battery when rpm is lowered and the generators cut out.

(4) **CARBURETOR AIR-HEAT.**

*(a)* Always land with the carburetor air-heat "ON." Beside ice prevention, carburetor heat is necessary to aid vaporization of fuel in the event

it is necessary to continue around without landing. The landing approach may cool the engines enough so that they will not pick up in an emergency because of poor fuel vaporization. Keep the carburetor air temperature below 40°C to avoid loss of power due to excessive heat.

*(b)* Open the cowl flaps immediately after landing. Taxi the airplane to the line with carburetor heat "ON." Temperature inversions are common in winter in arctic regions and the ground air may be 15°C (27°F) to 30°C (54°F) colder than at altitude. Therefore, care must be taken to avoid rapid cooling when letting down. If possible, maintain cylinder-head temperatures above 120°C.

(5) **BRAKES.**—While on the down-wind leg of the approach, work the pedals several times to insure circulation of sluggish fluid.

**h. POST-FLIGHT PROCEDURE.**

(1) **PARKING.**—Park the airplane on canvas or pine boughs to prevent the tires from freezing to the ground.

(2) **MOORING.**—In cold climates, where high winds are frequent, the airplanes, if parked outside, must be moored.

(3) **COVERS.**—Cover all surfaces which collect snow and ice, including the tail booms.

(4) **PARKING BRAKES.**— Do not set the parking brakes. Use wheel chocks.

**FLIGHT OPERATION INSTRUCTION CHART.**

—This chart includes information concerning range attainable and recommended power plant control settings for various combinations of gross weight, fuel load, altitude, and airspeed. To avoid misuse or misinterpretation of the chart, cognizance should be taken of these items:

1. The charted ranges make no allowance for warm-up, take-off, and climb. Fuel consumed during these operations should be obtained from Take-off, Climb, and Landing Chart. Similarly, no account is taken of the improved miles per gallon realizable during descent. Neglect of this latter factor is recommended to balance the fuel required for the landing operation.

2. The operating data included on any one chart should be used only when the gross weight is within the limits specified in the title block. When diminishing fuel load causes the gross weight to decrease to a value included in the weight limits of the next chart, the operating data included in the corresponding column of that chart should be used. THIS IS ESSENTIAL, AS RANGES HAVE BEEN COMPUTED ON THIS BASIS. Ranges shown on charts listing external load items are based on the conservative assumption that these items are carried the entire distance. If it is desirable to determine the range available if the auxiliary fuel tanks are dropped (in pairs) as emptied, this may be accomplished by planning the flight as a number of shorter flights, using the amount of fuel available under each external-load configuration, and referring to the appropriate charts for each segment of the flight.

3. All data have been based on the maximum weight for which the chart is applicable. When gross weight is within the chart weight limits and less than the maximum (due to lighter initial weight or diminished fuel load), the airspeed should be slightly greater than that listed on the chart. To be conservative, no account has been taken of this factor.

4. Experience has indicated that it is necessary to reduce calculated (red) range data by 15 per cent and reliable flight test (black) range data by 5 per cent to take account of variations in service airplanes and operating techniques. These allowances have been made on the Flight Operation Instruction Chart by a corresponding increase in fuel consumption. NO ALLOWANCE HAS BEEN MADE FOR WIND, NAVIGATIONAL ERROR, OR OTHER CONTINGENCIES. NO ALLOWANCE HAS BEEN MADE FOR COMBAT OR FORMATION FLIGHT. APPROPRIATE ALLOWANCES FOR THESE ITEMS SHOULD BE DICTATED BY LOCAL DOCTRINE. The fuel quantity entered in using the chart, therefore, should be the fuel available after reaching flight altitude, less allowances appropriate for the mission.

5. The range quoted depends upon steady operation at conditions given in the corresponding operating data. If the flight consists of a number of periods of operation under widely different conditions, it can be planned as a number of short flights.

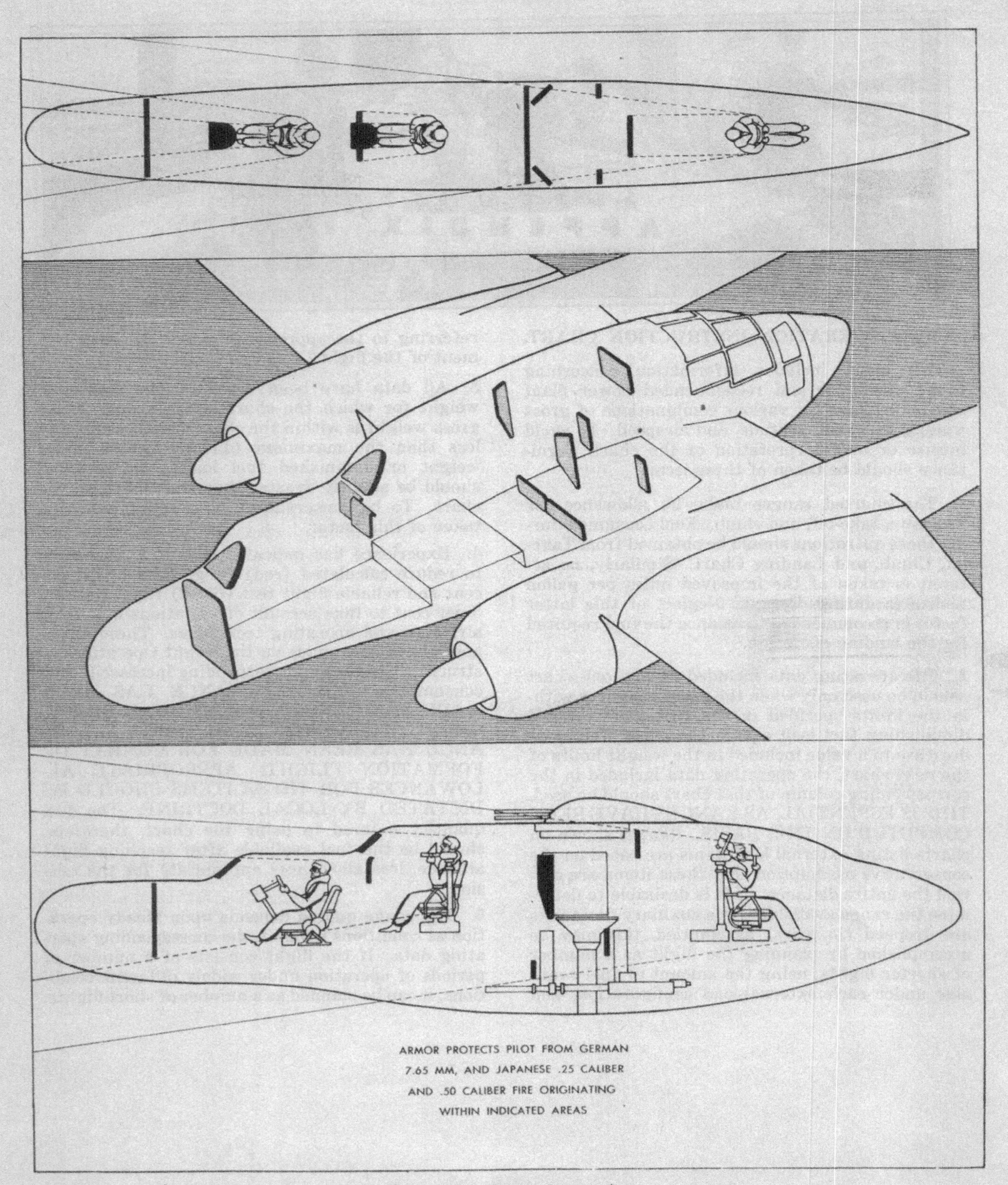


***Figure 65 — Armor Protection***

AAFMC-527
4-1-44

AIRCRAFT MODEL(S)
P-61C

# TAKE-OFF, CLIMB & LANDING CHART

ENGINE MODEL(S)
(2) R-2800-73

## TAKE-OFF DISTANCE FEET

| GROSS WEIGHT LBS | HEAD WIND | | HARD SURFACE RUNWAY | | | | | | SOD-TURF RUNWAY | | | | | | SOFT SURFACE RUNWAY | | | | | |
|---|---|---|---|---|---|---|---|---|---|---|---|---|---|---|---|---|---|---|---|---|
| | | | AT SEA LEVEL | | AT 3000 FEET | | AT 6000 FEET | | AT SEA LEVEL | | AT 3000 FEET | | AT 6000 FEET | | AT SEA LEVEL | | AT 3000 FEET | | AT 6000 FEET | |
| | MPH | KTS | GROUND RUN | TO CLEAR 50' OBJ. | GROUND RUN | TO CLEAR 50' OBJ. | GROUND RUN | TO CLEAR 50' OBJ. | GROUND RUN | TO CLEAR 50' OBJ. | GROUND RUN | TO CLEAR 50' OBJ. | GROUND RUN | TO CLEAR 50' OBJ. | GROUND RUN | TO CLEAR 50' OBJ. | GROUND RUN | TO CLEAR 50' OBJ. | GROUND RUN | TO CLEAR 50' OBJ. |
| 40,000 | 0 | 0 | | | | | | | | | | | | | | | | | | |
| | 20 | 17 | | | | | | | | | | | | | | | | | | |
| | 40 | 35 | | | | | | | | | | | | | | | | | | |
| 35,000 | 0 | 0 | | | | | | | | | | | | | | | | | | |
| | 20 | 17 | | | | | | | | | | | | | | | | | | |
| | 40 | 35 | | | | | | | | | | | | | | | | | | |
| 30,000 | 0 | 0 | | | | | | | | | | | | | | | | | | |
| | 20 | 17 | | | | | | | | | | | | | | | | | | |
| | 40 | 35 | | | | | | | | | | | | | | | | | | |

NOTE: INCREASE CHART DISTANCES AS FOLLOWS: 75 F - 10%; 100 F - 20%; 125 F - 30%; 150 F - 40% OPTIMUM TAKE-OFF WITH 2800 RPM 54.0 IN. Hg & 20 DEG FLAP IS 80% OF CHART VALUES
DATA AS OF 4-1-45 BASED ON: ESTIMATED DATA

## CLIMB DATA

| GROSS WEIGHT LBS | AT SEA LEVEL | | | | AT 5000 FEET | | | | | AT 10,000 FEET | | | | | AT 15,000 FEET | | | | | AT 25,000 FEET | | | | | AT 35,000 FEET | | | | |
|---|---|---|---|---|---|---|---|---|---|---|---|---|---|---|---|---|---|---|---|---|---|---|---|---|---|---|---|---|---|
| | BEST IAS | | RATE OF CLIMB | GAL OF FUEL | BEST IAS | | RATE OF CLIMB | FROM SL | | BEST IAS | | RATE OF CLIMB | FROM SL | | BEST IAS | | RATE OF CLIMB | FROM SL | | BEST IAS | | RATE OF CLIMB | FROM SL | | BEST IAS | | RATE OF CLIMB | FROM SL | |
| | MPH | KTS | FPM | USED | MPH | KTS | FPM | TIME MIN | FUEL USED | MPH | KTS | FPM | TIME MIN | FUEL USED | MPH | KTS | FPM | TIME MIN | FUEL USED | MPH | KTS | FPM | TIME MIN | FUEL USED | MPH | KTS | FPM | TIME MIN | FUEL USED |
| 40,000 | | | | | | | | | | | | | | | | | | | | | | | | | | | | | |
| 35,000 | | | | | | | | | | | | | | | | | | | | | | | | | | | | | |
| 30,000 | | | | | | | | | | | | | | | | | | | | | | | | | | | | | |

POWER PLANT SETTINGS: (DETAILS ON FIG. 32, SECTION III)
DATA AS OF 4-1-45 BASED ON: ESTIMATED DATA FUEL USED (US GAL) INCLUDES WARM-UP & TAKE-OFF ALLOWANCE

## LANDING DISTANCE FEET

| GROSS WEIGHT LBS | BEST IAS APPROACH | | | | HARD DRY SURFACE | | | | | | FIRM DRY SOD | | | | | | WET OR SLIPPERY | | | | | |
|---|---|---|---|---|---|---|---|---|---|---|---|---|---|---|---|---|---|---|---|---|---|---|
| | POWER OFF | | POWER ON | | AT SEA LEVEL | | AT 3000 FEET | | AT 6000 FEET | | AT SEA LEVEL | | AT 3000 FEET | | AT 6000 FEET | | AT SEA LEVEL | | AT 3000 FEET | | AT 6000 FEET | |
| | MPH | KTS | MPH | KTS | GROUND ROLL | TO CLEAR 50' OBJ. | GROUND ROLL | TO CLEAR 50' OBJ. | GROUND ROLL | TO CLEAR 50' OBJ. | GROUND ROLL | TO CLEAR 50' OBJ. | GROUND ROLL | TO CLEAR 50' OBJ. | GROUND ROLL | TO CLEAR 50' OBJ. | GROUND ROLL | TO CLEAR 50' OBJ. | GROUND ROLL | TO CLEAR 50' OBJ. | GROUND ROLL | TO CLEAR 50' OBJ. |
| 30,000 | 125 | | 115 | | | | | | | | | | | | | | | | | | | |
| 25,000 | 115 | | 105 | | 1400 | 2050 | | | | | | | | | | | | | | | | |

DATA AS OF 4-1-45 BASED ON: ESTIMATED DATA AND XP-61 FLIGHT TESTS OPTIMUM LANDING IS 80% OF CHART VALUES

**REMARKS:**

NOTE: TO DETERMINE FUEL CONSUMPTION IN BRITISH IMPERIAL GALLONS, MULTIPLY BY 10, THEN DIVIDE BY 12

**LEGEND**

IAS: Indicated Airspeed
MPH: Miles Per Hour
KTS: Knots
FPM: Feet Per Minute

***Figure 66 — Take-off, Climb, and Landing Chart***

| AAFMC-528 4-1-44 | AIRCRAFT MODEL(S) P-61C ENGINE(S): (2) R-2800-73 | FLIGHT OPERATION INSTRUCTION CHART CHART WEIGHT LIMITS: 32,000 TO 29,000 POUNDS | EXTERNAL LOAD ITEMS NONE NUMBER OF ENGINES OPERATING: TWO |
|---|---|---|---|

| LIMITS | RPM | MP (IN. Hg) | BLOWER POSITION | MIXTURE POSITION | TIME LIMIT | CYL. TEMP | TOTAL GPH | FOR DETAILS SEE POWER PLANT CHART (FIG. 32, SECT. III) |
|---|---|---|---|---|---|---|---|---|
| WAR EMERG. | 2800 | | | AR | 5 Min | 260 | | |
| MILITARY POWER | 2800 | | | AR | 15 Min | 260 | | |

INSTRUCTIONS FOR USING CHART: Select figure in FUEL column equal to or less than amount of fuel to be used for cruising. [1] Move horizontally to right or left and select RANGE value equal to or greater than the statute or nautical air miles to be flown. Vertically below and opposite value nearest desired cruising altitude (ALT) read RPM, manifold pressure (MP) and MIXTURE setting required.

NOTES: Column I is for emergency high speed cruising only. Columns II, III, IV and V give progressive increase in range at a sacrifice in speed. Air miles per gallon (MI/GAL) (no wind), gallons per hr. (GPH) and true airspeed (TAS) are approximate values for reference. Range values are for an average airplane flying alone (no wind). [1] To obtain BRITISH IMPERIAL GAL. (or GPH): multiply US gal (or GPH) by 10 then divide by 12.

| COLUMN I | | FUEL | COLUMN II | | COLUMN III | | COLUMN IV | | FUEL | COLUMN V | |
|---|---|---|---|---|---|---|---|---|---|---|---|
| RANGE IN AIR MILES | | US | RANGE IN AIR MILES | | RANGE IN AIR MILES | | RANGE IN AIR MILES | | US | RANGE IN AIR MILES | |
| STATUTE | NAUTICAL | GAL | STATUTE | NAUTICAL | STATUTE | NAUTICAL | STATUTE | NAUTICAL | GAL | STATUTE | NAUTICAL |
| | | 640<br>600<br>500 | SUBTRACT FUEL ALLOWANCES NOT AVAILABLE FOR CRUISING [1] | | | | | | 640<br>600<br>500 | | |
| | | 400<br>300<br>200 | | | | | | | 400<br>300<br>200 | | |
| | | 100 | | | | | | | 100 | | |
| | | | | | | | | | | | |

| MAXIMUM CONTINUOUS | | | | | | Press | (.80 STAT (.69 NAUT) MI/GAL) | | | | | |
|---|---|---|---|---|---|---|---|---|---|---|---|---|
| RPM | MP IN. | MIX-TURE | APPROX TOT GPH | TAS MPH | TAS KTS | ALT Feet | RPM | MP IN. | MIX-TURE | APPROX TOT GPH | TAS MPH | TAS KTS |
| | | | | | | 40000 | | | | | | |
| 2600 | | AR | | | | 35000 | | | | | | |
| 2600 | | AR | | | | 30000 | 2600 | 41.5 | AL | | | |
| 2600 | | AR | | | | 25000 | 2600 | 41.5 | AL | | | |
| 2600 | | AR | | | | 20000 | 2600 | 41.5 | AL | | | |
| 2600 | | AR | | | | 15000 | 2550 | 39.5 | AL | | | |
| 2600 | | AR | | | | 10000 | 2500 | 37.5 | AL | | | |
| 2600 | | AR | | | | 5000 | 2450 | 37.0 | AL | | | |
| 2600 | | AR | | | | SL | 2400 | 36.5 | AL | | | |

| (1.00 STAT (.87 NAUT) MI/GAL) | | | | | | (1.20 STAT (1.04 NAUT) MI/GAL) | | | | | | Press | MAXIMUM AIR RANGE | | | | | |
|---|---|---|---|---|---|---|---|---|---|---|---|---|---|---|---|---|---|---|
| RPM | MP IN. | MIX-TURE | APPROX TOT GPH | TAS MPH | TAS KTS | RPM | MP IN. | MIX-TURE | APPROX TOT GPH | TAS MPH | TAS KTS | ALT Feet | RPM | MP IN. | MIX-TURE | APPROX TOT GPH | TAS MPH | TAS KTS |
| | | | | | | | | | | | | 40000 | | | | | | |
| | | | | | | | | | | | | 35000 | | | | | | |
| 2400 | 34.0 | AL | | | | | | | | | | 30000 | | | | | | |
| 2400 | 34.0 | AL | | | | 2200 | 32.5 | AL | | | | 25000 | | | | | | |
| 2400 | 34.0 | AL | | | | 2200 | 32.5 | AL | | | | 20000 | | | | | | |
| 2400 | 34.0 | AL | | | | 2150 | 32.5 | AL | | | | 15000 | | | | | | |
| 2300 | 33.5 | AL | | | | 2100 | 33.0 | AL | | | | 10000 | 1650 | 34.5 | AL | | | |
| 2300 | 34.5 | AL | | | | 2000 | 34.0 | AL | | | | 5000 | 1550 | 35.0 | AL | | | |
| 2200 | 35.0 | AL | | | | 1900 | 35.5 | AL | | | | SL | 1400 | 35.0 | AL | | | |

SPECIAL NOTES

(1) MAKE ALLOWANCE FOR WARM-UP, TAKE-OFF & CLIMB (SEE FIGURE 66) PLUS ALLOWANCE FOR WIND, RESERVE AND COMBAT AS REQUIRED.

EXAMPLE

AT 32,000 LB GROSS WEIGHT WITH 430 GAL OF FUEL (AFTER DEDUCTING TOTAL ALLOWANCES OF 210 GAL) TO FLY 400 STAT AIR MILES AT 20,000 FT ALTITUDE MAINTAIN 2400 RPM AND 34.0 IN. MANIFOLD PRESSURE WITH MIXTURE SET: "AUTO LEAN"

LEGEND

ALT: PRESSURE ALTITUDE
MP: MANIFOLD PRESSURE
GPH: US GAL PER HR
TAS: TRUE AIRSPEED
KTS: KNOTS
SL: SEA LEVEL
FR: FULL RICH
AR: AUTO-RICH
AL: AUTO-LEAN
CL: CRUISING LEAN
ML: MANUAL LEAN
FT: FULL THROTTLE

DATA AS OF 6-15-45 BASED ON: CALCULATIONS

RED FIGURES ARE PRELIMINARY DATA, SUBJECT TO REVISION AFTER FLIGHT CHECK

**Figure 67 — Flight Operating Instruction Chart (Sheet 1 of 2 Sheets)**

| AAFMC-528 4-1-44 | AIRCRAFT MODEL(S) P-61C ENGINE(S): (2) R-2800-73 | FLIGHT OPERATION INSTRUCTION CHART CHART WEIGHT LIMITS: 29,000 TO 25,000 POUNDS | EXTERNAL LOAD ITEMS NONE NUMBER OF ENGINES OPERATING: TWO |
|---|---|---|---|

| LIMITS | RPM | MP IN. Hg | BLOWER POSITION | MIXTURE POSITION | TIME LIMIT | CYL TEMP | TOTAL GPH |
|---|---|---|---|---|---|---|---|
| WAR EMERG. | 2800 | | | AR | 5 Min | 260 | |
| MILITARY POWER | 2800 | | | AR | 15 Min | 260 | |

FOR DETAILS SEE POWER PLANT CHART (FIG. 32, SECT. III)

INSTRUCTIONS FOR USING CHART: Select figure in FUEL column equal to or less than amount of fuel to be used for cruising. [1] Move horizontally to right or left and select RANGE value equal to or greater than the statute or nautical air miles to be flown. Vertically below and opposite value nearest desired cruising altitude (ALT) read RPM, manifold pressure (MP) and MIXTURE setting required.

NOTES: Column I is for emergency high speed cruising only. Columns II, III, IV and V give progressive increase in range at a sacrifice in speed. Air miles per gallon (MI/GAL) (no wind), gallons per hr. (GPH) and true airspeed (TAS) are approximate values for reference. Range values are for an average airplane flying alone (no wind). [1] To obtain BRITISH IMPERIAL GAL. (or GPH): multiply US gal (or GPH) by 10 then divide by 12.

| COLUMN I | | FUEL | COLUMN II | | COLUMN III | | COLUMN IV | | FUEL | COLUMN V | |
|---|---|---|---|---|---|---|---|---|---|---|---|
| RANGE IN AIR MILES | | US | RANGE IN AIR MILES | | RANGE IN AIR MILES | | RANGE IN AIR MILES | | US | RANGE IN AIR MILES | |
| STATUTE | NAUTICAL | GAL | STATUTE | NAUTICAL | STATUTE | NAUTICAL | STATUTE | NAUTICAL | GAL | STATUTE | NAUTICAL |
| | | 640<br>600<br>500 | SUBTRACT FUEL ALLOWANCES NOT AVAILABLE FOR CRUISING [1] | | | | | | 640<br>600<br>500 | | |
| | | 400<br>300<br>200 | | | | | | | 400<br>300<br>200 | | |
| | | 100 | | | | | | | 100 | | |
| | | | | | | | | | | | |
| | | | | | | | | | | | |

| MAXIMUM CONTINUOUS | | | | | | Press | (.85 STAT (.74 NAUT) MI/GAL) | | | | | | (1.05 STAT (.91 NAUT) MI/GAL) | | | | | | (1.35 STAT (1.17 NAUT) MI/GAL) | | | | | | Press | MAXIMUM AIR RANGE | | | | | |
|---|---|---|---|---|---|---|---|---|---|---|---|---|---|---|---|---|---|---|---|---|---|---|---|---|---|---|---|---|---|---|---|
| RPM | MP IN. | MIXTURE | APPROX TOT GPH | APPROX TAS MPH | APPROX TAS KTS | ALT Feet | RPM | MP IN. | MIXTURE | APPROX TOT GPH | APPROX TAS MPH | APPROX TAS KTS | RPM | MP IN. | MIXTURE | APPROX TOT GPH | APPROX TAS MPH | APPROX TAS KTS | RPM | MP IN. | MIXTURE | APPROX TOT GPH | APPROX TAS MPH | APPROX TAS KTS | ALT Feet | RPM | MP IN. | MIXTURE | APPROX TOT GPH | APPROX TAS MPH | APPROX TAS KTS |
| | | | | | | 40000 | | | | | | | | | | | | | | | | | | | 40000 | | | | | | |
| 2600 | | AR | | | | 35000 | | | | | | | | | | | | | | | | | | | 35000 | | | | | | |
| 2600 | | AR | | | | 30000 | 2600 | 41.5 | AL | | | | 2400 | 34.0 | AL | | | | | | | | | | 30000 | | | | | | |
| 2600 | | AR | | | | 25000 | 2550 | 39.5 | AL | | | | 2400 | 34.0 | AL | | | | | | | | | | 25000 | | | | | | |
| 2600 | | AR | | | | 20000 | 2500 | 37.5 | AL | | | | 2400 | 34.0 | AL | | | | | | | | | | 20000 | | | | | | |
| 2600 | | AR | | | | 15000 | 2500 | 37.5 | AL | | | | 2350 | 33.5 | AL | | | | 1900 | 33.0 | AL | | | | 15000 | | | | | | |
| 2600 | | AR | | | | 10000 | 2450 | 36.5 | AL | | | | 2300 | 33.5 | AL | | | | 1900 | 33.0 | AL | | | | 10000 | 1600 | 29.5 | AL | | | |
| 2600 | | AR | | | | 5000 | 2450 | 37.0 | AL | | | | 2250 | 34.0 | AL | | | | 1800 | 34.0 | AL | | | | 5000 | 1500 | 31.5 | AL | | | |
| 2600 | | AR | | | | SL | 2400 | 36.5 | AL | | | | 2250 | 35.5 | AL | | | | 1700 | 35.0 | AL | | | | SL | 1400 | 32.0 | AL | | | |

**SPECIAL NOTES**

(1) MAKE ALLOWANCE FOR WARM-UP, TAKE-OFF & CLIMB (SEE FIGURE 66) PLUS ALLOWANCE FOR WIND, RESERVE AND COMBAT AS REQUIRED.

USE FUEL FROM INE'D DROP TANKS FIRST.

**EXAMPLE**

AT 29,000 LB GROSS WEIGHT WITH 560 GAL OF FUEL (AFTER DEDUCTING TOTAL ALLOWANCES OF 180 GAL) TO FLY 500 STAT AIR MILES AT 20,000 FT ALTITUDE MAINTAIN 2400 RPM AND 34.0 IN. MANIFOLD PRESSURE WITH MIXTURE SET: "AUTO LEAN"

**LEGEND**

ALT: PRESSURE ALTITUDE
MP: MANIFOLD PRESSURE
GPH: US GAL PER HR
TAS: TRUE AIRSPEED
KTS: KNOTS
SL: SEA LEVEL
FR: FULL RICH
AR: AUTO-RICH
AL: AUTO-LEAN
CL: CRUISING LEAN
ML: MANUAL LEAN
FT: FULL THROTTLE

DATA AS OF 6-15-45 BASED ON: CALCULATIONS

RED FIGURES ARE PRELIMINARY DATA, SUBJECT TO REVISION AFTER FLIGHT CHECK

*Figure 67 — Flight Operating Instruction Chart (Sheet 2 of 2 Sheets)*

AAFMC-528
4-1-44

AIRCRAFT MODEL(S): P-61C
ENGINE(S): (2) R-2800-73

# FLIGHT OPERATION INSTRUCTION CHART

CHART WEIGHT LIMITS: 35,000 TO 32,000 POUNDS

EXTERNAL LOAD ITEMS
**FOUR 165-GAL. DROP TANKS, OR TWO OUTB'D 310-GAL. DROP TANKS**
NUMBER OF ENGINES OPERATING: **TWO**

| LIMITS | RPM | MP IN. Hg | BLOWER POSITION | MIXTURE POSITION | TIME LIMIT | CYL TEMP | TOTAL GPH |
|---|---|---|---|---|---|---|---|
| WAR EMERG. | 2800 | | | AR | 5 Min | 260 | |
| MILITARY POWER | 2800 | | | AR | 15 Min | 260 | |

FOR DETAILS SEE POWER PLANT CHART (FIG. 32, SECT. III)

INSTRUCTIONS FOR USING CHART: Select figure in FUEL column equal to or less than amount of fuel to be used for cruising.[1] Move horizontally to right or left and select RANGE value equal to or greater than the statute or nautical air miles to be flown. Vertically below and opposite value nearest desired cruising altitude (ALT) read RPM, manifold pressure (MP) and MIXTURE setting required.

NOTES: Column I is for emergency high speed cruising only. Columns II, III, IV and V give progressive increase in range at a sacrifice in speed. Air miles per gallon (MI/GAL) (no wind), gallons per hr. (GPH) and true airspeed (TAS) are approximate values for reference. Range values are for an average airplane flying alone (no wind).[1] To obtain BRITISH IMPERIAL GAL. (or GPH): multiply US gal (or GPH) by 10 then divide by 12.

| COLUMN I | | FUEL | COLUMN II | | COLUMN III | | COLUMN IV | | FUEL | COLUMN V | |
|---|---|---|---|---|---|---|---|---|---|---|---|
| RANGE IN AIR MILES | | US | RANGE IN AIR MILES | | RANGE IN AIR MILES | | RANGE IN AIR MILES | | US | RANGE IN AIR MILES | |
| STATUTE | NAUTICAL | GAL | STATUTE | NAUTICAL | STATUTE | NAUTICAL | STATUTE | NAUTICAL | GAL | STATUTE | NAUTICAL |
| | | 1300 | | | | | | | 1300 | | |
| | | 1200 | | | | | | | 1200 | | |
| | | 1100 | | | | | | | 1100 | | |
| | | 1000 | | | | | | | 1000 | | |
| | | 900 | | | | | | | 900 | | |
| | | 800 | | | | | | | 800 | | |
| | | 700 | | | | | | | 700 | | |
| | | 600 | | | | | | | 600 | | |
| | | 500 | | | | | | | 500 | | |
| | | 400 | | | | | | | 400 | | |
| | | 300 | | | | | | | 300 | | |
| | | 200 | | | | | | | 200 | | |

SUBTRACT FUEL ALLOWANCES NOT AVAILABLE FOR CRUISING [1]

| Press ALT Feet | MAXIMUM CONTINUOUS RPM | MP IN. | MIXTURE | APPROX TOT GPH | TAS MPH | TAS KTS | (.75 STAT (.65 NAUT) MI/GAL) RPM | MP IN. | MIXTURE | APPROX TOT GPH | TAS MPH | TAS KTS |
|---|---|---|---|---|---|---|---|---|---|---|---|---|
| 40000 | | | | | | | | | | | | |
| 35000 | | | | | | | | | | | | |
| 30000 | 2600 | | AR | | | | | | | | | |
| 25000 | 2600 | | AR | | | | 2550 | 39.5 | AL | | | |
| 20000 | 2600 | | AR | | | | 2550 | 39.5 | AL | | | |
| 15000 | 2600 | | AR | | | | 2500 | 37.5 | AL | | | |
| 10000 | 2600 | | AR | | | | 2500 | 38.0 | AL | | | |
| 5000 | 2600 | | AR | | | | 2450 | 37.0 | AL | | | |
| SL | 2600 | | AR | | | | 2450 | 38.0 | AL | | | |

| Press ALT Feet | (.90 STAT (.78 NAUT) MI/GAL) RPM | MP IN. | MIXTURE | APPROX TOT GPH | TAS MPH | TAS KTS | (1.00 STAT (.87 NAUT) MI/GAL) RPM | MP IN. | MIXTURE | APPROX TOT GPH | TAS MPH | TAS KTS |
|---|---|---|---|---|---|---|---|---|---|---|---|---|
| 40000 | | | | | | | | | | | | |
| 35000 | | | | | | | | | | | | |
| 30000 | | | | | | | | | | | | |
| 25000 | 2400 | 34.0 | AL | | | | | | | | | |
| 20000 | 2400 | 34.0 | AL | | | | 2250 | 33.0 | AL | | | |
| 15000 | 2400 | 34.0 | AL | | | | 2250 | 33.0 | AL | | | |
| 10000 | 2350 | 34.0 | AL | | | | 2250 | 33.0 | AL | | | |
| 5000 | 2300 | 34.5 | AL | | | | 2200 | 34.0 | AL | | | |
| SL | 2250 | 35.5 | AL | | | | 2100 | 35.5 | AL | | | |

| Press ALT Feet | MAXIMUM AIR RANGE RPM | MP IN. | MIXTURE | APPROX TOT GPH | TAS MPH | TAS KTS |
|---|---|---|---|---|---|---|
| 40000 | | | | | | |
| 35000 | | | | | | |
| 30000 | | | | | | |
| 25000 | | | | | | |
| 20000 | | | | | | |
| 15000 | | | | | | |
| 10000 | 1900 | 33.0 | AL | | | |
| 5000 | 1800 | 34.0 | AL | | | |
| SL | 1700 | 35.0 | AL | | | |

**SPECIAL NOTES**

(1) MAKE ALLOWANCE FOR WARM-UP, TAKE-OFF & CLIMB (SEE FIGURE 66) PLUS ALLOWANCE FOR WIND, RESERVE AND COMBAT AS REQUIRED.

IF FOUR DROP TANKS ARE CARRIED, USE FUEL FROM INB'D DROP TANKS FIRST.

**EXAMPLE**

AT 35,000 LB GROSS WEIGHT WITH 795 GAL OF FUEL (AFTER DEDUCTING TOTAL ALLOWANCES OF 305 GAL) TO FLY 600 STAT AIR MILES AT 20,000 FT ALTITUDE MAINTAIN 2400 RPM AND 34.0 IN. MANIFOLD PRESSURE WITH MIXTURE SET: "AUTO LEAN"

**LEGEND**

ALT: PRESSURE ALTITUDE
MP: MANIFOLD PRESSURE
GPH: US GAL PER HR
TAS: TRUE AIRSPEED
KTS: KNOTS
SL: SEA LEVEL
FR: FULL RICH
AR: AUTO-RICH
AL: AUTO-LEAN
CL: CRUISING LEAN
ML: MANUAL LEAN
FT: FULL THROTTLE

DATA AS OF 6-15-45 BASED ON: CALCULATIONS

RED FIGURES ARE PRELIMINARY DATA, SUBJECT TO REVISION AFTER FLIGHT CHECK

*Figure 68 — Flight Operating Instruction Chart (Sheet 1 of 3 Sheets)*

| AAFMC-528 4-1-44 | AIRCRAFT MODEL(S) P-61C ENGINE(S): (2) R-2800-73 | FLIGHT OPERATION INSTRUCTION CHART CHART WEIGHT LIMITS: 32,000 TO 29,000 POUNDS | EXTERNAL LOAD ITEMS FOUR 165-GAL. DROP TANKS, OR TWO OUTB'D 310-GAL. DROP TANKS NUMBER OF ENGINES OPERATING: TWO |
|---|---|---|---|

| LIMITS | RPM | MP IN. Hg | BLOWER POSITION | MIXTURE POSITION | TIME LIMIT | CYL. TEMP | TOTAL GPH |
|---|---|---|---|---|---|---|---|
| WAR EMERG. | 2800 | | | AR | 5 Min | 260 | |
| MILITARY POWER | 2800 | | | AR | 15 Min | 260 | |

[1] FOR DETAILS SEE POWER PLANT CHART (FIG. 33, SECT. III)

INSTRUCTIONS FOR USING CHART: Select figure in FUEL column equal to or less than amount of fuel to be used for cruising. [1] Move horizontally to right or left and select RANGE value equal to or greater than the statute or nautical air miles to be flown. Vertically below and opposite value nearest desired cruising altitude (ALT) read RPM, manifold pressure (MP) and MIXTURE setting required.

NOTES: Column I is for emergency high speed cruising only. Columns II, III, IV and V give progressive increase in range at a sacrifice in speed. Air miles per gallon (MI/GAL) (no wind), gallons per hr. (GPH) and true airspeed (TAS) are approximate values for reference. Range values are for an average airplane flying alone (no wind). [1] To obtain BRITISH IMPERIAL GAL. (or GPH): multiply US gal (or GPH) by 10 then divide by 12.

| COLUMN I | | FUEL | COLUMN II | | COLUMN III | | COLUMN IV | | FUEL | COLUMN V | |
|---|---|---|---|---|---|---|---|---|---|---|---|
| RANGE IN AIR MILES | | US | RANGE IN AIR MILES | | RANGE IN AIR MILES | | RANGE IN AIR MILES | | US | RANGE IN AIR MILES | |
| STATUTE | NAUTICAL | GAL | STATUTE | NAUTICAL | STATUTE | NAUTICAL | STATUTE | NAUTICAL | GAL | STATUTE | NAUTICAL |
| | | 1100<br>1000<br>900 | SUBTRACT FUEL ALLOWANCES NOT AVAILABLE FOR CRUISING [1] | | | | | | 1100<br>1000<br>900 | | |
| | | 800<br>700<br>600 | | | | | | | 800<br>700<br>600 | | |
| | | 500<br>400<br>300 | | | | | | | 500<br>400<br>300 | | |
| | | 200<br>100 | | | | | | | 200<br>100 | | |

| MAXIMUM CONTINUOUS | | | | | | Press | (.80 STAT (.69 NAUT) MI/GAL) | | | | | |
|---|---|---|---|---|---|---|---|---|---|---|---|---|
| RPM | MP IN. | MIXTURE | APPROX TOT GPH | TAS MPH | TAS KTS | ALT Feet | RPM | MP IN. | MIXTURE | APPROX TOT GPH | TAS MPH | TAS KTS |
| | | | | | | 40000 | | | | | | |
| 2600 | | AR | | | | 35000 | | | | | | |
| 2600 | | AR | | | | 30000 | 2550 | 39.5 | AL | | | |
| 2600 | | AR | | | | 25000 | 2550 | 39.5 | AL | | | |
| 2600 | | AR | | | | 20000 | 2500 | 37.5 | AL | | | |
| 2600 | | AR | | | | 15000 | 2500 | 37.5 | AL | | | |
| 2600 | | AR | | | | 10000 | 2450 | 36.5 | AL | | | |
| 2600 | | AR | | | | 5000 | 2400 | 35.5 | AL | | | |
| 2600 | | AR | | | | SL | 2400 | 36.5 | AL | | | |

| Press | (.95 STAT (.83 NAUT) MI/GAL) | | | | | | (1.10 STAT (.95 NAUT) MI/GAL) | | | | | |
|---|---|---|---|---|---|---|---|---|---|---|---|---|
| ALT Feet | RPM | MP IN. | MIXTURE | APPROX TOT GPH | TAS MPH | TAS KTS | RPM | MP IN. | MIXTURE | APPROX TOT GPH | TAS MPH | TAS KTS |
| 40000 | | | | | | | | | | | | |
| 35000 | | | | | | | | | | | | |
| 30000 | | | | | | | | | | | | |
| 25000 | 2400 | 34.0 | AL | | | | | | | | | |
| 20000 | 2400 | 34.0 | AL | | | | 2200 | 32.0 | AL | | | |
| 15000 | 2350 | 33.5 | AL | | | | 2200 | 32.0 | AL | | | |
| 10000 | 2300 | 33.5 | AL | | | | 2100 | 33.0 | AL | | | |
| 5000 | 2300 | 34.5 | AL | | | | 2000 | 34.0 | AL | | | |
| SL | 2200 | 35.0 | AL | | | | 2000 | 35.5 | AL | | | |

| Press | MAXIMUM AIR RANGE | | | | | |
|---|---|---|---|---|---|---|
| ALT Feet | RPM | MP IN. | MIXTURE | APPROX TOT GPH | TAS MPH | TAS KTS |
| 40000 | | | | | | |
| 35000 | | | | | | |
| 30000 | | | | | | |
| 25000 | | | | | | |
| 20000 | | | | | | |
| 15000 | | | | | | |
| 10000 | 1800 | 33.5 | AL | | | |
| 5000 | 1700 | 33.5 | AL | | | |
| SL | 1500 | 34.5 | AL | | | |

SPECIAL NOTES

(1) MAKE ALLOWANCE FOR WARM-UP, TAKE-OFF & CLIMB (SEE FIGURE 66) PLUS ALLOWANCE FOR WIND, RESERVE AND COMBAT AS REQUIRED.

IF FOUR DROP TANKS ARE CARRIED, USE FUEL FROM INB'D DROP TANKS FIRST.

EXAMPLE

AT 32,000 LB GROSS WEIGHT WITH 890 GAL OF FUEL (AFTER DEDUCTING TOTAL ALLOWANCES OF 210 GAL) TO FLY 750 STAT AIR MILES AT 20,000 FT ALTITUDE MAINTAIN 2400 RPM AND 34.0 IN. MANIFOLD PRESSURE WITH MIXTURE SET: "AUTO LEAN"

LEGEND

ALT: PRESSURE ALTITUDE
MP: MANIFOLD PRESSURE
GPH: US GAL PER HR
TAS: TRUE AIRSPEED
KTS: KNOTS
SL: SEA LEVEL
FR: FULL RICH
AR: AUTO-RICH
AL: AUTO-LEAN
CL: CRUISING LEAN
ML: MANUAL LEAN
FT: FULL THROTTLE

DATA AS OF 6-15-45 BASED ON: CALCULATIONS

RED FIGURES ARE PRELIMINARY DATA, SUBJECT TO REVISION AFTER FLIGHT CHECK

Figure 68 — Flight Operating Instruction Chart (Sheet 2 of 3 Sheets)

AAFMC-528
4-1-44

AIRCRAFT MODEL(S): P-61C
ENGINE(S): (2) R-2800-73

# FLIGHT OPERATION INSTRUCTION CHART

CHART WEIGHT LIMITS: 29,000 TO 25,000 POUNDS

EXTERNAL LOAD ITEMS
**FOUR 165-GAL. DROP TANKS, OR TWO OUTB'D 310-GAL. DROP TANKS**

NUMBER OF ENGINES OPERATING: **TWO**

| LIMITS | RPM | MP (IN. Hg) | BLOWER POSITION | MIXTURE POSITION | TIME LIMIT | CYL TEMP | TOTAL GPH |
|---|---|---|---|---|---|---|---|
| WAR EMERG. | 2800 | | | AR | 5 Min | 260 | |
| MILITARY POWER | 2800 | | | AR | 15 Min | 260 | |

FOR DETAILS SEE POWER PLANT CHART (FIG. 32, SECT. III)

**INSTRUCTIONS FOR USING CHART:** Select figure in FUEL column equal to or less than amount of fuel to be used for cruising. [1] Move horizontally to right or left and select RANGE value equal to or greater than the statute or nautical air miles to be flown. Vertically below and opposite value nearest desired cruising altitude (ALT) read RPM, manifold pressure (MP) and MIXTURE setting required.

**NOTES:** Column I is for emergency high speed cruising only. Columns II, III, IV and V give progressive increase in range at a sacrifice in speed. Air miles per gallon (MI/GAL) (no wind), gallons per hr. (GPH) and true airspeed (TAS) are approximate values for reference. Range values are for an average airplane flying alone (no wind). [1] To obtain BRITISH IMPERIAL GAL. (or GPH): multiply US gal (or GPH) by 10 then divide by 12.

| COLUMN I | | FUEL | COLUMN II | | COLUMN III | | COLUMN IV | | FUEL | COLUMN V | |
|---|---|---|---|---|---|---|---|---|---|---|---|
| RANGE IN AIR MILES | | US | RANGE IN AIR MILES | | RANGE IN AIR MILES | | RANGE IN AIR MILES | | US | RANGE IN AIR MILES | |
| STATUTE | NAUTICAL | GAL | STATUTE | NAUTICAL | STATUTE | NAUTICAL | STATUTE | NAUTICAL | GAL | STATUTE | NAUTICAL |
| | | 640 | SUBTRACT FUEL ALLOWANCES NOT AVAILABLE FOR CRUISING [1] | | | | | | 640 | | |
| | | 600 | | | | | | | 600 | | |
| | | 500 | | | | | | | 500 | | |
| | | 400 | | | | | | | 400 | | |
| | | 300 | | | | | | | 300 | | |
| | | 200 | | | | | | | 200 | | |
| | | 100 | | | | | | | 100 | | |

| Press ALT Feet | MAXIMUM CONTINUOUS RPM | MP IN. | MIX-TURE | APPROX TOT GPH | TAS MPH | TAS KTS | COLUMN II (.85 STAT (.74 NAUT) MI/GAL) RPM | MP IN. | MIX-TURE | APPROX TOT GPH | TAS MPH | TAS KTS |
|---|---|---|---|---|---|---|---|---|---|---|---|---|
| 40000 | | | | | | | | | | | | |
| 35000 | 2600 | | AR | | | | 2550 | 39.5 | AL | | | |
| 30000 | 2600 | | AR | | | | 2500 | 37.5 | AL | | | |
| 25000 | 2600 | | AR | | | | 2500 | 37.5 | AL | | | |
| 20000 | 2600 | | AR | | | | 2450 | 36.0 | AL | | | |
| 15000 | 2600 | | AR | | | | 2450 | 36.0 | AL | | | |
| 10000 | 2600 | | AR | | | | 2400 | 34.5 | AL | | | |
| 5000 | 2600 | | AR | | | | 2400 | 35.5 | AL | | | |
| SL | 2600 | | AR | | | | 2350 | 36.5 | AL | | | |

| Press ALT Feet | COLUMN III (1.05 STAT (.91 NAUT) MI/GAL) RPM | MP IN. | MIX-TURE | APPROX TOT GPH | TAS MPH | TAS KTS | COLUMN IV (1.20 STAT (1.04 NAUT) MI/GAL) RPM | MP IN. | MIX-TURE | APPROX TOT GPH | TAS MPH | TAS KTS | MAXIMUM AIR RANGE RPM | MP IN. | MIX-TURE | APPROX TOT GPH | TAS MPH | TAS KTS |
|---|---|---|---|---|---|---|---|---|---|---|---|---|---|---|---|---|---|---|
| 40000 | | | | | | | | | | | | | | | | | | |
| 35000 | | | | | | | | | | | | | | | | | | |
| 30000 | 2350 | 33.5 | AL | | | | | | | | | | | | | | | |
| 25000 | 2350 | 33.5 | AL | | | | | | | | | | | | | | | |
| 20000 | 2300 | 33.5 | AL | | | | 2100 | 33.0 | AL | | | | | | | | | |
| 15000 | 2250 | 33.0 | AL | | | | 2100 | 33.0 | AL | | | | | | | | | |
| 10000 | 2250 | 33.0 | AL | | | | 2000 | 33.5 | AL | | | | 1650 | 34.5 | AL | | | |
| 5000 | 2200 | 34.0 | AL | | | | 1900 | 34.0 | AL | | | | 1600 | 34.5 | AL | | | |
| SL | 2100 | 35.5 | AL | | | | 1800 | 35.5 | AL | | | | 1400 | 35.0 | AL | | | |

**SPECIAL NOTES**

(1) MAKE ALLOWANCE FOR WARM-UP, TAKE-OFF & CLIMB (SEE FIGURE 66) PLUS ALLOWANCE FOR WIND, RESERVE AND COMBAT AS REQUIRED.

IF FOUR DROP TANKS ARE CARRIED, USE FUEL FROM INB'D DROP TANKS FIRST.

**EXAMPLE**

AT 29,000 LB GROSS WEIGHT WITH 560 GAL OF FUEL (AFTER DEDUCTING TOTAL ALLOWANCES OF 180 GAL) TO FLY 500 STAT AIR MILES AT 20,000 FT ALTITUDE MAINTAIN 2300 RPM AND 33.5 IN. MANIFOLD PRESSURE WITH MIXTURE SET: "AUTO LEAN"

**LEGEND**

ALT: PRESSURE ALTITUDE
MP: MANIFOLD PRESSURE
GPH: US GAL PER HR
TAS: TRUE AIRSPEED
KTS: KNOTS
SL: SEA LEVEL
FR: FULL RICH
AR: AUTO-RICH
AL: AUTO-LEAN
CL: CRUISING LEAN
ML: MANUAL LEAN
FT: FULL THROTTLE

DATA AS OF 6-15-45 BASED ON: CALCULATIONS

RED FIGURES ARE PRELIMINARY DATA, SUBJECT TO REVISION AFTER FLIGHT CHECK

*Figure 68 — Flight Operating Instruction Chart (Sheet 3 of 3 Sheets)*

AAFMC-528 4-1-44

| AIRCRAFT MODEL(S) | FLIGHT OPERATION INSTRUCTION CHART | EXTERNAL LOAD ITEMS |
|---|---|---|
| P-61C | CHART WEIGHT LIMITS: **40,000** TO **35,000** POUNDS | FOUR 310-GAL. DROP TANKS |
| ENGINE(S): (2) R-2800-73 | | NUMBER OF ENGINES OPERATING: TWO |

| LIMITS | RPM | MP (IN. Hg) | BLOWER POSITION | MIXTURE POSITION | TIME LIMIT | CYL TEMP | TOTAL GPH |
|---|---|---|---|---|---|---|---|
| WAR EMERG. | 2800 | | | AR | 5 Min | 260 | |
| MILITARY POWER | 2800 | | | AR | 15 Min | 260 | |

FOR DETAILS SEE POWER PLANT CHART (FIG. 32, SECT. III)

INSTRUCTIONS FOR USING CHART: Select figure in FUEL column equal to or less than amount of fuel to be used for cruising. [1] Move horizontally to right or left and select RANGE value equal to or greater than the statute or nautical air miles to be flown. Vertically below and opposite value nearest desired cruising altitude (ALT) read RPM, manifold pressure (MP) and MIXTURE setting required.

NOTES: Column I is for emergency high speed cruising only. Columns II, III, IV and V give progressive increase in range at a sacrifice in speed. Air miles per gallon (MI/GAL) (no wind), gallons per hr. (GPH) and true airspeed (TAS) are approximate values for reference. Range values are for an average airplane flying alone (no wind). [1] To obtain BRITISH IMPERIAL GAL (or GPH): multiply US gal (or GPH) by 10 then divide by 12.

| COLUMN I | | FUEL | COLUMN II | | COLUMN III | | COLUMN IV | | FUEL | COLUMN V | |
|---|---|---|---|---|---|---|---|---|---|---|---|
| RANGE IN AIR MILES | | US | RANGE IN AIR MILES | | RANGE IN AIR MILES | | RANGE IN AIR MILES | | US | RANGE IN AIR MILES | |
| STATUTE | NAUTICAL | GAL | STATUTE | NAUTICAL | STATUTE | NAUTICAL | STATUTE | NAUTICAL | GAL | STATUTE | NAUTICAL |
| | | | SUBTRACT FUEL ALLOWANCES NOT AVAILABLE FOR CRUISING [1] | | | | | | | | |
| | | 1880 | | | | | | | 1880 | | |
| | | 1800 | | | | | | | 1800 | | |
| | | 1700 | | | | | | | 1700 | | |
| | | 1600 | | | | | | | 1600 | | |
| | | 1500 | | | | | | | 1500 | | |
| | | 1400 | | | | | | | 1400 | | |
| | | 1300 | | | | | | | 1300 | | |
| | | 1200 | | | | | | | 1200 | | |
| | | 1100 | | | | | | | 1100 | | |
| | | 1000 | | | | | | | 1000 | | |
| | | 900 | | | | | | | 900 | | |
| | | 800 | | | | | | | 800 | | |

| MAXIMUM CONTINUOUS | | | | | | Press | (.65 STAT (.56 NAUT) MI/GAL) | | | | | | (.65 STAT (.56 NAUT) MI/GAL) | | | | | | (.75 STAT (.65 NAUT) MI/GAL) | | | | | | Press | MAXIMUM AIR RANGE | | | | | |
|---|---|---|---|---|---|---|---|---|---|---|---|---|---|---|---|---|---|---|---|---|---|---|---|---|---|---|---|---|---|---|---|
| RPM | MP IN. | MIXTURE | APPROX TOT GPH | TAS MPH | TAS KTS | ALT Feet | RPM | MP IN. | MIXTURE | APPROX TOT GPH | TAS MPH | TAS KTS | RPM | MP IN. | MIXTURE | APPROX TOT GPH | TAS MPH | TAS KTS | RPM | MP IN. | MIXTURE | APPROX TOT GPH | TAS MPH | TAS KTS | ALT Feet | RPM | MP IN. | MIXTURE | APPROX TOT GPH | TAS MPH | TAS KTS |
| | | | | | | 40000 | | | | | | | | | | | | | | | | | | | 40000 | | | | | | |
| | | | | | | 35000 | | | | | | | | | | | | | | | | | | | 35000 | | | | | | |
| | | | | | | 30000 | | | | | | | | | | | | | | | | | | | 30000 | | | | | | |
| 2600 | | AR | | | | 25000 | | | | | | | | | | | | | | | | | | | 25000 | | | | | | |
| 2600 | | AR | | | | 20000 | 2500 | 37.5 | AL | | | | | | | | | | | | | | | | 20000 | | | | | | |
| 2600 | | AR | | | | 15000 | 2500 | 37.5 | AL | | | | Use Operating Data shown | | | | | | | | | | | | 15000 | | | | | | |
| 2600 | | AR | | | | 10000 | 2500 | 38.0 | AL | | | | in Col. II for this weight | | | | | | 2400 | 34.5 | AL | | | | 10000 | 2300 | 33.5 | AL | | | |
| 2600 | | AR | | | | 5000 | 2450 | 37.0 | AL | | | | range only. Use Col. III on | | | | | | 2400 | 35.5 | AL | | | | 5000 | 2250 | 34.0 | AL | | | |
| 2600 | | AR | | | | SL | 2450 | 38.0 | AL | | | | lower weight band charts. | | | | | | 2350 | 36.5 | AL | | | | SL | 2100 | 36.0 | AL | | | |

SPECIAL NOTES

(1) MAKE ALLOWANCE FOR WARM-UP, TAKE-OFF & CLIMB (SEE FIGURE 66) PLUS ALLOWANCE FOR WIND, RESERVE AND COMBAT AS REQUIRED.

USE FUEL FROM INB'D DROP TANKS FIRST.

EXAMPLE

AT **40,000** LB GROSS WEIGHT WITH **1520** GAL OF FUEL (AFTER DEDUCTING TOTAL ALLOWANCES OF **360** GAL) TO FLY **1100** STAT AIR MILES AT **20,000** FT ALTITUDE MAINTAIN **2500** RPM AND **37.5** IN. MANIFOLD PRESSURE WITH MIXTURE SET: **"AUTO LEAN"**

LEGEND

ALT: PRESSURE ALTITUDE
MP: MANIFOLD PRESSURE
GPH: US GAL PER HR
TAS: TRUE AIRSPEED
KTS: KNOTS
SL: SEA LEVEL
FR: FULL RICH
AR: AUTO-RICH
AL: AUTO-LEAN
CL: CRUISING LEAN
ML: MANUAL LEAN
FT: FULL THROTTLE

DATA AS OF 6-15-45 BASED ON: CALCULATIONS

RED FIGURES ARE PRELIMINARY DATA, SUBJECT TO REVISION AFTER FLIGHT CHECK

*Figure 69— Flight Operating Instruction Chart (Sheet 1 of 4 Sheets)*

AAFMC-528
4-1-44

AIRCRAFT MODEL(S): P-61C
ENGINE(S): (2) R-2800-73

# FLIGHT OPERATION INSTRUCTION CHART

CHART WEIGHT LIMITS: 35,000 TO 32,000 POUNDS

EXTERNAL LOAD ITEMS: FOUR 310-GAL. DROP TANKS

NUMBER OF ENGINES OPERATING: TWO

| LIMITS | RPM | MP IN. Hg | BLOWER POSITION | MIXTURE POSITION | TIME LIMIT | CYL TEMP | TOTAL GPH |
|---|---|---|---|---|---|---|---|
| WAR EMERG. | 2800 | | | AR | 5 Min | 260 | |
| MILITARY POWER | 2800 | | | AR | 15 Min | 260 | |

FOR DETAILS SEE POWER-PLANT CHART (FIG. 32, SECT. III)

INSTRUCTIONS FOR USING CHART: Select figure in FUEL column equal to or less than amount of fuel to be used for cruising. [1] Move horizontally to right or left and select RANGE value equal to or greater than the statute or nautical air miles to be flown. Vertically below and opposite value nearest desired cruising altitude (ALT) read RPM, manifold pressure (MP) and MIXTURE setting required.

NOTES: Column I is for emergency high speed cruising only. Columns II, III, IV and V give progressive increase in range at a sacrifice in speed. Air miles per gallon (MI/GAL) (no wind), gallons per hr. (GPH) and true airspeed (TAS) are approximate values for reference. Range values are for an average airplane flying alone (no wind). [1] To obtain BRITISH IMPERIAL GAL. (or GPH): multiply US gal (or GPH) by 10 then divide by 12.

| COLUMN I | | FUEL | COLUMN II | | COLUMN III | | COLUMN IV | | FUEL | COLUMN V | |
|---|---|---|---|---|---|---|---|---|---|---|---|
| RANGE IN AIR MILES | | US | RANGE IN AIR MILES | | RANGE IN AIR MILES | | RANGE IN AIR MILES | | US | RANGE IN AIR MILES | |
| STATUTE | NAUTICAL | GAL | STATUTE | NAUTICAL | STATUTE | NAUTICAL | STATUTE | NAUTICAL | GAL | STATUTE | NAUTICAL |
| | | | SUBTRACT FUEL ALLOWANCES NOT AVAILABLE FOR CRUISING [1] | | | | | | | | |
| | | 1500 | | | | | | | 1500 | | |
| | | 1400 | | | | | | | 1400 | | |
| | | 1300 | | | | | | | 1300 | | |
| | | 1200 | | | | | | | 1200 | | |
| | | 1100 | | | | | | | 1100 | | |
| | | 1000 | | | | | | | 1000 | | |
| | | 900 | | | | | | | 900 | | |
| | | 800 | | | | | | | 800 | | |
| | | 700 | | | | | | | 700 | | |
| | | 600 | | | | | | | 600 | | |
| | | 500 | | | | | | | 500 | | |

| MAXIMUM CONTINUOUS | | | | | | Press | (.70 STAT (.61 NAUT) MI/GAL) | | | | | | (.80 STAT (.69 NAUT) MI/GAL) | | | | | | (.90 STAT (.78 NAUT) MI/GAL) | | | | | | Press | MAXIMUM AIR RANGE | | | | | |
|---|---|---|---|---|---|---|---|---|---|---|---|---|---|---|---|---|---|---|---|---|---|---|---|---|---|---|---|---|---|---|---|
| RPM | MP IN. | MIX-TURE | APPROX TOT GPH | TAS MPH | TAS KTS | ALT Feet | RPM | MP IN. | MIX-TURE | APPROX TOT GPH | TAS MPH | TAS KTS | RPM | MP IN. | MIX-TURE | APPROX TOT GPH | TAS MPH | TAS KTS | RPM | MP IN. | MIX-TURE | APPROX TOT GPH | TAS MPH | TAS KTS | ALT Feet | RPM | MP IN. | MIX-TURE | APPROX TOT GPH | TAS MPH | TAS KTS |
| | | | | | | 40000 | | | | | | | | | | | | | | | | | | | 40000 | | | | | | |
| | | | | | | 35000 | | | | | | | | | | | | | | | | | | | 35000 | | | | | | |
| 2600 | | AR | | | | 30000 | | | | | | | | | | | | | | | | | | | 30000 | | | | | | |
| 2600 | | AR | | | | 25000 | 2550 | 39.5 | AL | | | | | | | | | | | | | | | | 25000 | | | | | | |
| 2600 | | AR | | | | 20000 | 2550 | 39.5 | AL | | | | 2400 | 34.0 | AL | | | | | | | | | | 20000 | | | | | | |
| 2600 | | AR | | | | 15000 | 2500 | 37.5 | AL | | | | 2400 | 34.0 | AL | | | | 2250 | 33.0 | AL | | | | 15000 | | | | | | |
| 2600 | | AR | | | | 10000 | 2500 | 38.0 | AL | | | | 2400 | 34.5 | AL | | | | 2250 | 33.0 | AL | | | | 10000 | 2000 | 33.5 | AL | | | |
| 2600 | | AR | | | | 5000 | 2450 | 37.0 | AL | | | | 2350 | 35.0 | AL | | | | 2200 | 34.0 | AL | | | | 5000 | 2000 | 34.0 | AL | | | |
| 2600 | | AR | | | | SL | 2450 | 38.0 | AL | | | | 2350 | 36.5 | AL | | | | 2200 | 35.0 | AL | | | | SL | 1800 | 35.5 | AL | | | |

SPECIAL NOTES

(1) MAKE ALLOWANCE FOR WARM-UP, TAKE-OFF & CLIMB (SEE FIGURE 66) PLUS ALLOWANCE FOR WIND, RESERVE AND COMBAT AS REQUIRED.

USE FUEL FROM INB'D DROP TANKS FIRST.

EXAMPLE

AT 35,000 LB GROSS WEIGHT WITH 1200 GAL OF FUEL (AFTER DEDUCTING TOTAL ALLOWANCES OF 305 GAL) TO FLY 1000 STAT AIR MILES AT 20,000 FT ALTITUDE MAINTAIN 2400 RPM AND 34.0 IN. MANIFOLD PRESSURE WITH MIXTURE SET: "AUTO LEAN"

LEGEND

ALT: PRESSURE ALTITUDE
MP: MANIFOLD PRESSURE
GPH: US GAL PER HR
TAS: TRUE AIRSPEED
KTS: KNOTS
SL: SEA LEVEL
FR: FULL RICH
AR: AUTO-RICH
AL: AUTO-LEAN
CL: CRUISING LEAN
ML: MANUAL LEAN
FT: FULL THROTTLE

DATA AS OF 6-15-45 BASED ON: CALCULATIONS

RED FIGURES ARE PRELIMINARY DATA, SUBJECT TO REVISION AFTER FLIGHT CHECK

*Figure 69 — Flight Operating Instruction Chart (Sheet 2 of 4 Sheets)*

AAFMC-528 4-1-44

AIRCRAFT MODEL(S): P-61C
ENGINE(S): (2) R-2800-73

# FLIGHT OPERATION INSTRUCTION CHART

CHART WEIGHT LIMITS: **32,000** TO **29,000** POUNDS

EXTERNAL LOAD ITEMS: **FOUR 310-GAL. DROP TANKS**

NUMBER OF ENGINES OPERATING: **TWO**

| LIMITS | RPM | MP IN. Hg | BLOWER POSITION | MIXTURE POSITION | TIME LIMIT | CYL. TEMP | TOTAL GPH |
|---|---|---|---|---|---|---|---|
| WAR EMERG. | 2800 | | | AR | 5 Min | 260 | |
| MILITARY POWER | 2800 | | | AR | 15 Min | 260 | |

FOR DETAILS SEE POWER PLANT CHART (FIG. 32, SECT. III)

INSTRUCTIONS FOR USING CHART: Select figure in FUEL column equal to or less than amount of fuel to be used for cruising. [1] Move horizontally to right or left and select RANGE value equal to or greater than the statute or nautical air miles to be flown. Vertically below and opposite value nearest desired cruising altitude (ALT) read RPM, manifold pressure (MP) and MIXTURE setting required.

NOTES: Column I is for emergency high speed cruising only. Columns II, III, IV and V give progressive increase in range at a sacrifice in speed. Air miles per gallon (MI/GAL) (no wind), gallons per hr. (GPH) and true airspeed (TAS) are approximate values for reference. Range values are for an average airplane flying alone (no wind). [1] To obtain BRITISH IMPERIAL GAL. (or GPH): multiply US gal (or GPH) by 10 then divide by 12.

SUBTRACT FUEL ALLOWANCES NOT AVAILABLE FOR CRUISING [1]

| COLUMN I: RANGE IN AIR MILES, STATUTE | COLUMN I: NAUTICAL | FUEL US GAL | COLUMN II: RANGE IN AIR MILES, STATUTE | COLUMN II: NAUTICAL | COLUMN III: RANGE IN AIR MILES, STATUTE | COLUMN III: NAUTICAL | COLUMN IV: RANGE IN AIR MILES, STATUTE | COLUMN IV: NAUTICAL | FUEL US GAL | COLUMN V: RANGE IN AIR MILES, STATUTE | COLUMN V: NAUTICAL |
|---|---|---|---|---|---|---|---|---|---|---|---|
| | | 1100 | | | | | | | 1100 | | |
| | | 1000 | | | | | | | 1000 | | |
| | | 900 | | | | | | | 900 | | |
| | | 800 | | | | | | | 800 | | |
| | | 700 | | | | | | | 700 | | |
| | | 600 | | | | | | | 600 | | |
| | | 500 | | | | | | | 500 | | |
| | | 400 | | | | | | | 400 | | |
| | | 300 | | | | | | | 300 | | |
| | | 200 | | | | | | | 200 | | |
| | | 100 | | | | | | | 100 | | |

| MAXIMUM CONTINUOUS | | | | | | Press | (.75 STAT (.65 NAUT) MI/GAL) | | | | | | (.90 STAT (.78 NAUT) MI/GAL) | | | | | | (1.00 STAT (.87 NAUT) MI/GAL) | | | | | | Press | MAXIMUM AIR RANGE | | | | | |
|---|---|---|---|---|---|---|---|---|---|---|---|---|---|---|---|---|---|---|---|---|---|---|---|---|---|---|---|---|---|---|---|
| RPM | MP IN. | MIXTURE | APPROX TOT GPH | APPROX TAS MPH | APPROX TAS KTS | ALT Feet | RPM | MP IN. | MIXTURE | APPROX TOT GPH | APPROX TAS MPH | APPROX TAS KTS | RPM | MP IN. | MIXTURE | APPROX TOT GPH | APPROX TAS MPH | APPROX TAS KTS | RPM | MP IN. | MIXTURE | APPROX TOT GPH | APPROX TAS MPH | APPROX TAS KTS | ALT Feet | RPM | MP IN. | MIXTURE | APPROX TOT GPH | APPROX TAS MPH | APPROX TAS KTS |
| | | | | | | 40000 | | | | | | | | | | | | | | | | | | | 40000 | | | | | | |
| | | | | | | 35000 | | | | | | | | | | | | | | | | | | | 35000 | | | | | | |
| 2600 | | AR | | | | 30000 | | | | | | | | | | | | | | | | | | | 30000 | | | | | | |
| 2600 | | AR | | | | 25000 | 2500 | 37.5 | AL | | | | | | | | | | | | | | | | 25000 | | | | | | |
| 2600 | | AR | | | | 20000 | 2500 | 37.5 | AL | | | | 2350 | 33.5 | AL | | | | | | | | | | 20000 | | | | | | |
| 2600 | | AR | | | | 15000 | 2450 | 36.0 | AL | | | | 2300 | 33.5 | AL | | | | 2200 | 32.0 | AL | | | | 15000 | | | | | | |
| 2600 | | AR | | | | 10000 | 2450 | 36.5 | AL | | | | 2300 | 33.5 | AL | | | | 2200 | 32.5 | AL | | | | 10000 | 1900 | 33.0 | AL | | | |
| 2600 | | AR | | | | 5000 | 2400 | 35.5 | AL | | | | 2250 | 34.0 | AL | | | | 2100 | 34.0 | AL | | | | 5000 | 1800 | 34.0 | AL | | | |
| 2600 | | AR | | | | SL | 2400 | 36.5 | AL | | | | 2200 | 35.0 | AL | | | | 2000 | 35.5 | AL | | | | SL | 1700 | 35.0 | AL | | | |

**SPECIAL NOTES**

(1) MAKE ALLOWANCE FOR WARM-UP, TAKE-OFF & CLIMB (SEE FIGURE 66) PLUS ALLOWANCE FOR WIND, RESERVE AND COMBAT AS REQUIRED.

USE FUEL FROM INB'D DROP TANKS FIRST.

**EXAMPLE**

AT **32,000** LB GROSS WEIGHT WITH **790** GAL OF FUEL (AFTER DEDUCTING TOTAL ALLOWANCES OF **210** GAL) TO FLY **600** STAT AIR MILES AT **20,000** FT ALTITUDE MAINTAIN **2350** RPM AND **33.5** IN. MANIFOLD PRESSURE WITH MIXTURE SET: **"AUTO LEAN"**

**LEGEND**

ALT: PRESSURE ALTITUDE
MP: MANIFOLD PRESSURE
GPH: US GAL PER HR
TAS: TRUE AIRSPEED
KTS: KNOTS
SL: SEA LEVEL
FR: FULL RICH
AR: AUTO-RICH
AL: AUTO-LEAN
CL: CRUISING LEAN
ML: MANUAL LEAN
FT: FULL THROTTLE

DATA AS OF 6-15-45 BASED ON: CALCULATIONS

RED FIGURES ARE PRELIMINARY DATA, SUBJECT TO REVISION AFTER FLIGHT CHECK

*Figure 69 — Flight Operating Instruction Chart (Sheet 3 of 4 Sheets)*

AAFMC-528 4-1-44

AIRCRAFT MODEL(S): P-61C

ENGINE(S): (2) R-2800-73

# FLIGHT OPERATION INSTRUCTION CHART

CHART WEIGHT LIMITS: 29,000 TO 25,000 POUNDS

EXTERNAL LOAD ITEMS: FOUR 310-GAL. DROP TANKS

NUMBER OF ENGINES OPERATING: TWO

| LIMITS | RPM | MP IN. Hg | BLOWER POSITION | MIXTURE POSITION | TIME LIMIT | CYL TEMP | TOTAL GPH |
|---|---|---|---|---|---|---|---|
| WAR EMERG. | 2800 | | | AR | 5 Min | 260 | |
| MILITARY POWER | 2800 | | | AR | 15 Min | 260 | |

FOR DETAILS SEE POWER PLANT CHART (FIG. 32, SECT. III)

INSTRUCTIONS FOR USING CHART: Select figure in FUEL column equal to or less than amount of fuel to be used for cruising. [1] Move horizontally to right or left and select RANGE value equal to or greater than the statute or nautical air miles to be flown. Vertically below and opposite value nearest desired cruising altitude (ALT) read RPM, manifold pressure (MP) and MIXTURE setting required.

NOTES: Column I is for emergency high speed cruising only. Columns II, III, IV and V give progressive increase in range at a sacrifice in speed. Air miles per gallon (MI/GAL) (no wind), gallons per hr. (GPH) and true airspeed (TAS) are approximate values for reference. Range values are for an average airplane flying alone (no wind). [1] To obtain BRITISH IMPERIAL GAL. (or GPH): multiply US gal (or GPH) by 10 then divide by 12.

SUBTRACT FUEL ALLOWANCES NOT AVAILABLE FOR CRUISING [1]

| COLUMN I | | FUEL | COLUMN II | | COLUMN III | | COLUMN IV | | FUEL | COLUMN V | |
|---|---|---|---|---|---|---|---|---|---|---|---|
| RANGE IN AIR MILES | | US | RANGE IN AIR MILES | | RANGE IN AIR MILES | | RANGE IN AIR MILES | | US | RANGE IN AIR MILES | |
| STATUTE | NAUTICAL | GAL | STATUTE | NAUTICAL | STATUTE | NAUTICAL | STATUTE | NAUTICAL | GAL | STATUTE | NAUTICAL |
| | | 640 | | | | | | | 640 | | |
| | | 600 | | | | | | | 600 | | |
| | | 500 | | | | | | | 500 | | |
| | | 400 | | | | | | | 400 | | |
| | | 300 | | | | | | | 300 | | |
| | | 200 | | | | | | | 200 | | |
| | | 100 | | | | | | | 100 | | |

| MAXIMUM CONTINUOUS | | | | | | Press | (.80 STAT (.69 NAUT) MI/GAL) | | | | | | (.95 STAT (.83 NAUT) MI/GAL) | | | | | | (1.05 STAT (.91 NAUT) MI/GAL) | | | | | | Press | MAXIMUM AIR RANGE | | | | | |
|---|---|---|---|---|---|---|---|---|---|---|---|---|---|---|---|---|---|---|---|---|---|---|---|---|---|---|---|---|---|---|---|
| RPM | MP IN. | MIXTURE | APPROX TOT GPH | TAS MPH | TAS KTS | ALT Feet | RPM | MP IN. | MIXTURE | APPROX TOT GPH | TAS MPH | TAS KTS | RPM | MP IN. | MIXTURE | APPROX TOT GPH | TAS MPH | TAS KTS | RPM | MP IN. | MIXTURE | APPROX TOT GPH | TAS MPH | TAS KTS | ALT Feet | RPM | MP IN. | MIXTURE | APPROX TOT GPH | TAS MPH | TAS KTS |
| | | | | | | 40000 | | | | | | | | | | | | | | | | | | | 40000 | | | | | | |
| 2600 | | AR | | | | 35000 | | | | | | | | | | | | | | | | | | | 35000 | | | | | | |
| 2600 | | AR | | | | 30000 | 2500 | 37.5 | AL | | | | | | | | | | | | | | | | 30000 | | | | | | |
| 2600 | | AR | | | | 25000 | 2500 | 37.5 | AL | | | | 2350 | 33.5 | AL | | | | | | | | | | 25000 | | | | | | |
| 2600 | | AR | | | | 20000 | 2450 | 36.0 | AL | | | | 2300 | 33.5 | AL | | | | 2200 | 32.0 | AL | | | | 20000 | | | | | | |
| 2600 | | AR | | | | 15000 | 2450 | 36.0 | AL | | | | 2300 | 33.5 | AL | | | | 2200 | 32.0 | AL | | | | 15000 | | | | | | |
| 2600 | | AR | | | | 10000 | 2400 | 34.5 | AL | | | | 2250 | 33.0 | AL | | | | 2200 | 32.5 | AL | | | | 10000 | 1800 | 33.5 | AL | | | |
| 2600 | | AR | | | | 5000 | 2400 | 35.5 | AL | | | | 2200 | 34.0 | AL | | | | 2000 | 34.0 | AL | | | | 5000 | 1700 | 33.5 | AL | | | |
| 2600 | | AR | | | | SL | 2350 | 36.5 | AL | | | | 2100 | 35.5 | AL | | | | 1900 | 35.5 | AL | | | | SL | 1600 | 35.0 | AL | | | |

**SPECIAL NOTES**

(1) MAKE ALLOWANCE FOR WARM-UP, TAKE-OFF & CLIMB (SEE FIGURE 66) PLUS ALLOWANCE FOR WIND, RESERVE AND COMBAT AS REQUIRED.

USE FUEL FROM INB'D DROP TANKS FIRST

**EXAMPLE**

AT 29,000 LB GROSS WEIGHT WITH 560 GAL OF FUEL (AFTER DEDUCTING TOTAL ALLOWANCES OF 180 GAL) TO FLY 450 STAT AIR MILES AT 20,000 FT ALTITUDE MAINTAIN 2300 RPM AND 33.5 IN. MANIFOLD PRESSURE WITH MIXTURE SET: "AUTO LEAN"

**LEGEND**

ALT: PRESSURE ALTITUDE
MP: MANIFOLD PRESSURE
GPH: US GAL PER HR
TAS: TRUE AIRSPEED
KTS: KNOTS
SL: SEA LEVEL
FR: FULL RICH
AR: AUTO-RICH
AL: AUTO-LEAN
CL: CRUISING LEAN
ML: MANUAL LEAN
FT: FULL THROTTLE

DATA AS OF 6-15-45 BASED ON: CALCULATIONS

RED FIGURES ARE PRELIMINARY DATA, SUBJECT TO REVISION AFTER FLIGHT CHECK

Figure 69 — Flight Operating Instruction Chart (Sheet 4 of 4 Sheets)

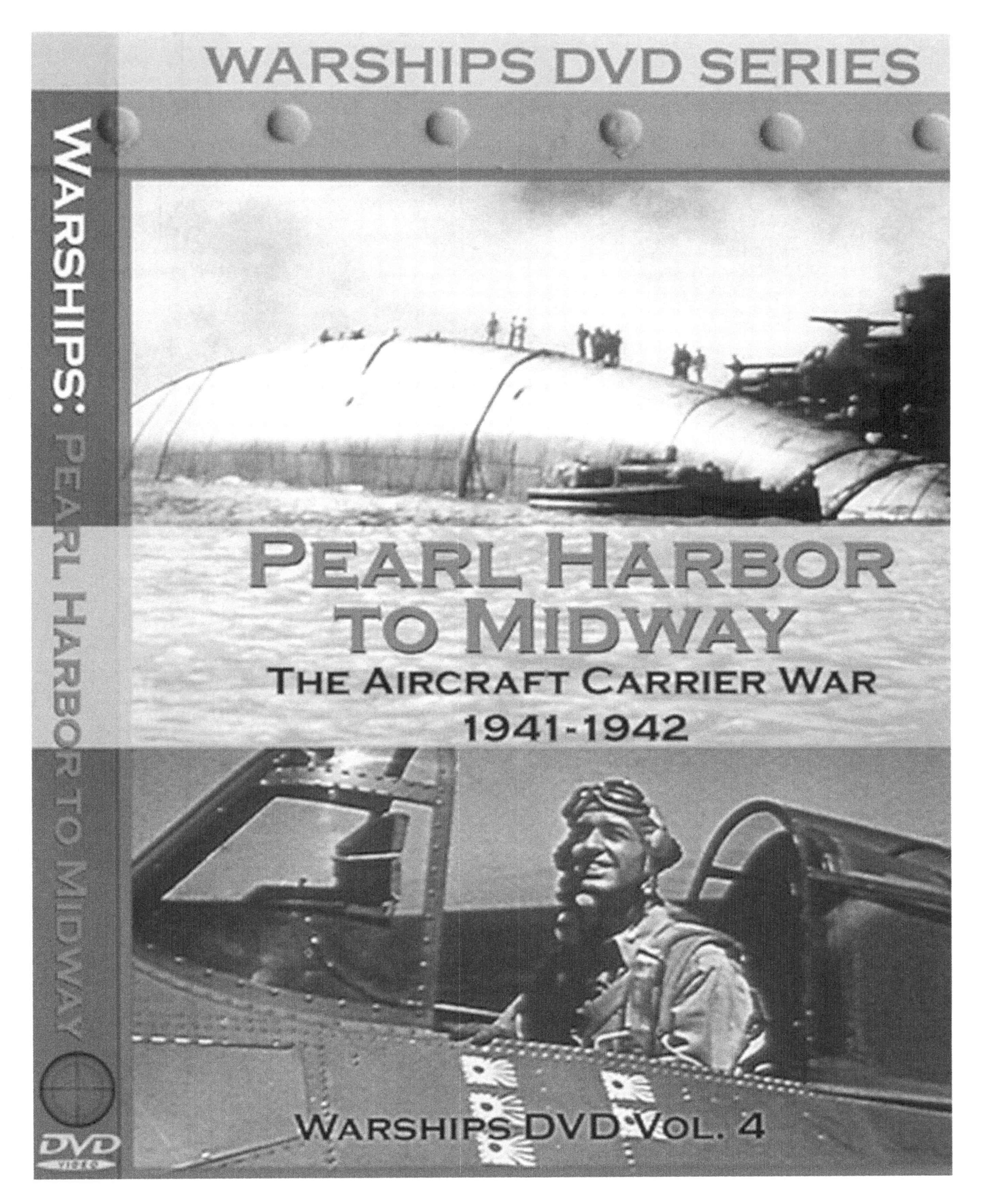


# NOW AVAILABLE!

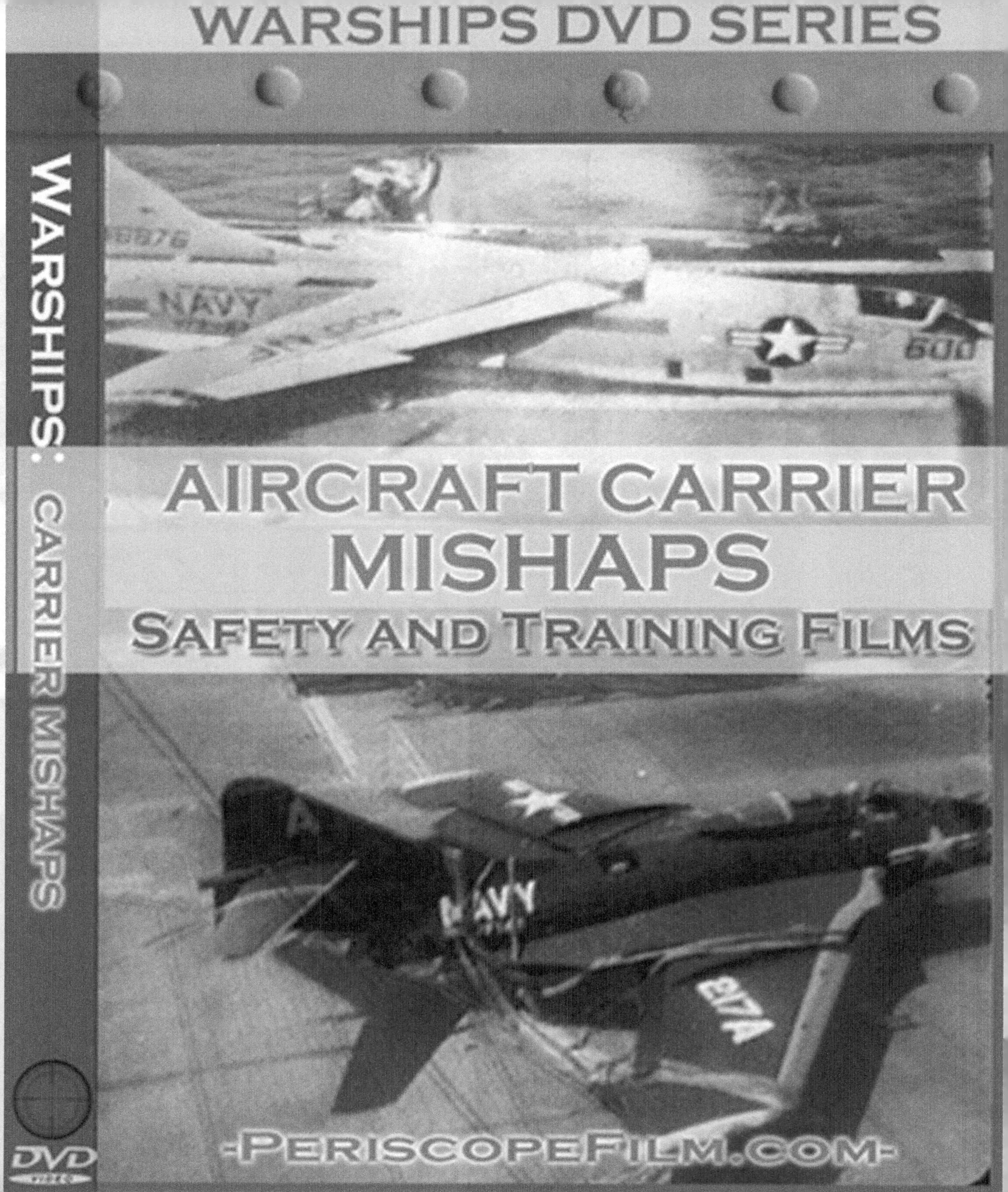
WARSHIPS DVD SERIES
WARSHIPS: CARRIER MISHAPS
AIRCRAFT CARRIER
MISHAPS
SAFETY AND TRAINING FILMS
NAVY
NAVY
-PERISCOPEFILM.COM-
DVD

# EPIC BATTLES OF WWII

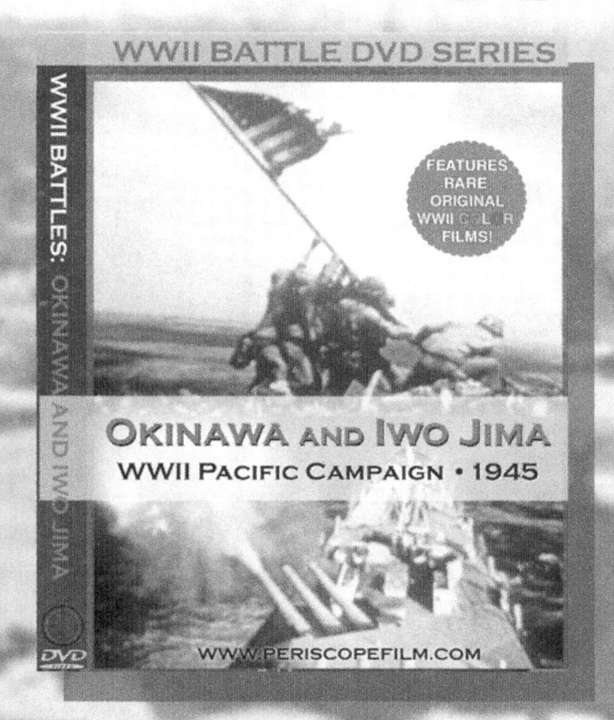


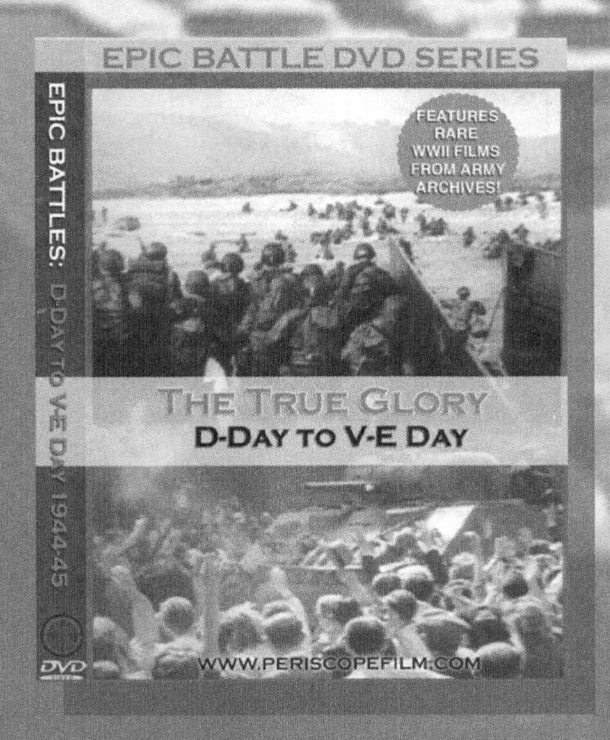

## NOW AVAILABLE ON DVD!

# Aircraft At War DVD Series

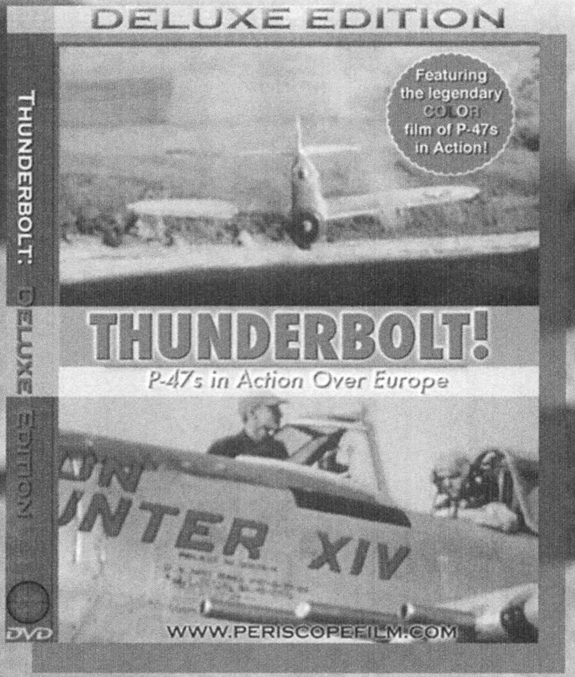


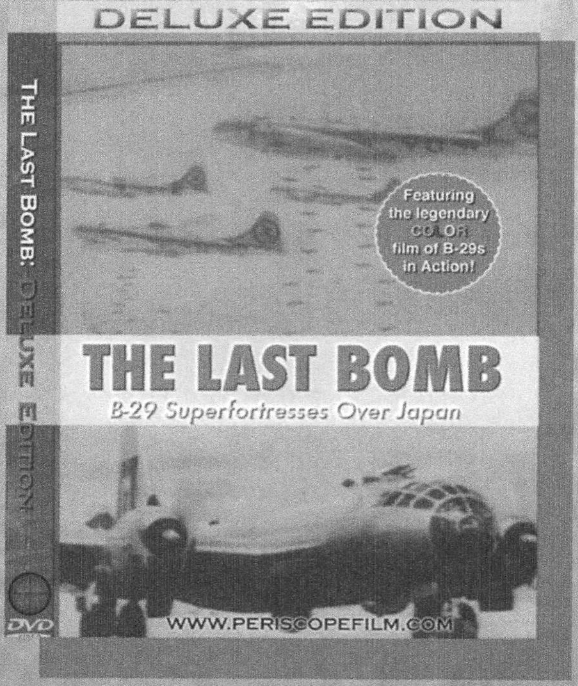


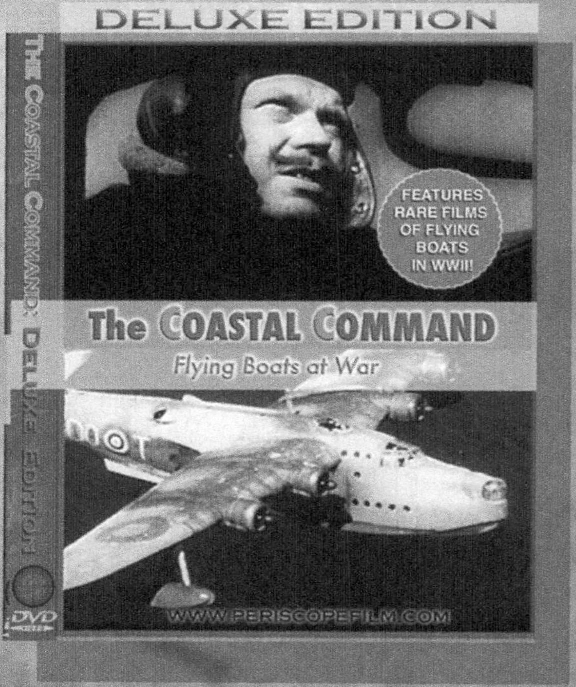


## Now Available!

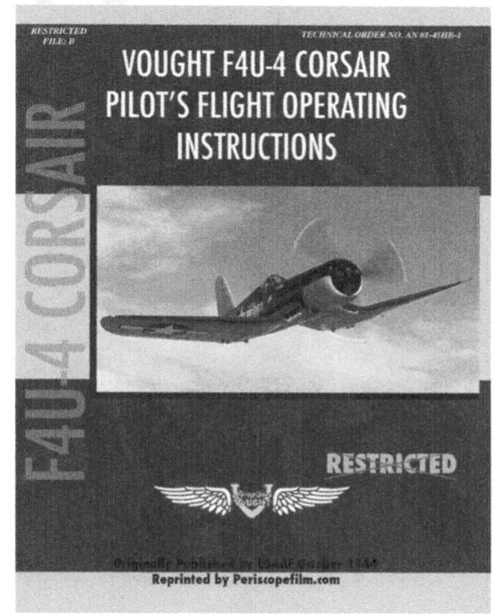
RESTRICTED FILE: B
VOUGHT F4U-4 CORSAIR PILOT'S FLIGHT OPERATING INSTRUCTIONS
F4U-4 CORSAIR
RESTRICTED
Reprinted by Periscopefilm.com

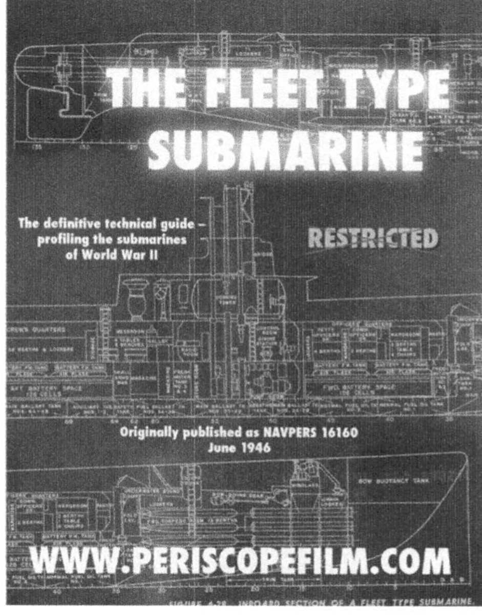
THE FLEET TYPE SUBMARINE
The definitive technical guide – profiling the submarines of World War II
RESTRICTED
Originally published as NAVPERS 16160 June 1946
WWW.PERISCOPEFILM.COM

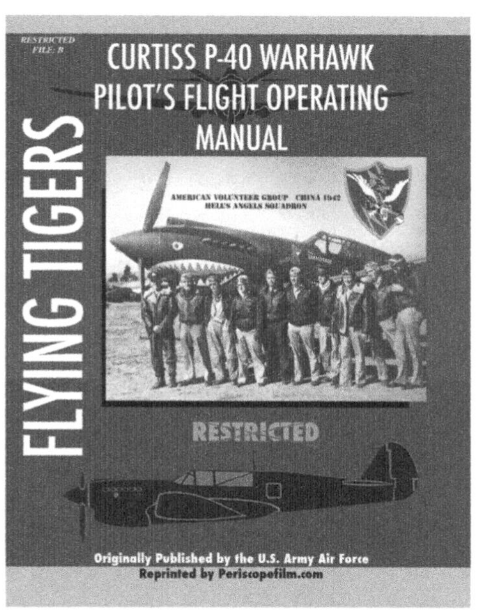
RESTRICTED FILE: B
CURTISS P-40 WARHAWK PILOT'S FLIGHT OPERATING MANUAL
FLYING TIGERS
RESTRICTED
Originally Published by the U.S. Army Air Force
Reprinted by Periscopefilm.com

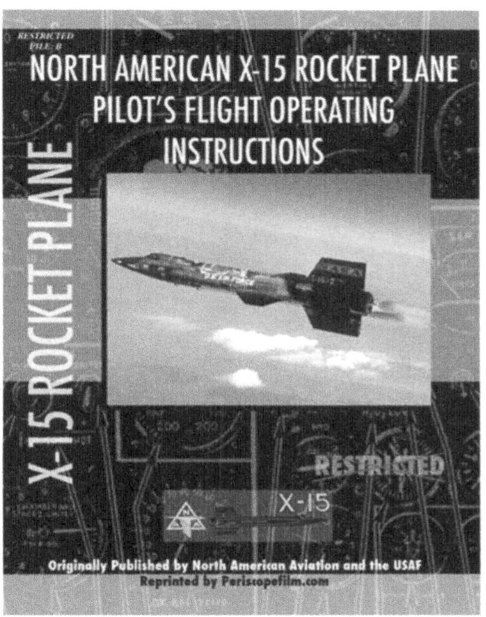
RESTRICTED FILE: B
NORTH AMERICAN X-15 ROCKET PLANE PILOT'S FLIGHT OPERATING INSTRUCTIONS
X-15 ROCKET PLANE
RESTRICTED
X-15
Originally Published by North American Aviation and the USAF
Reprinted by Periscopefilm.com

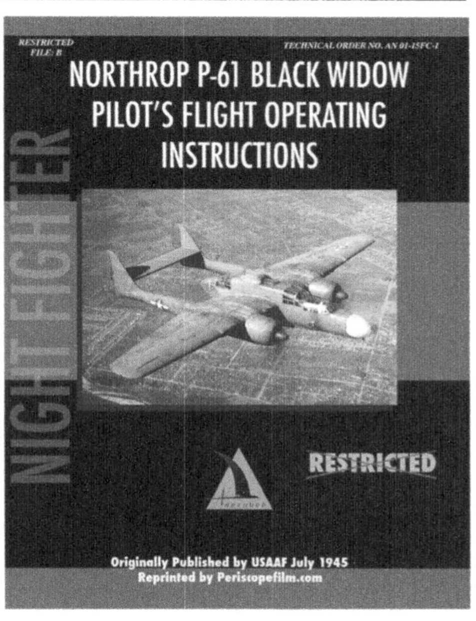
RESTRICTED FILE: B
NORTHROP P-61 BLACK WIDOW PILOT'S FLIGHT OPERATING INSTRUCTIONS
NIGHT FIGHTER
RESTRICTED
Originally Published by USAAF July 1945
Reprinted by Periscopefilm.com

RESTRICTED FILE: B
NORTHROP YB-49 FLYING WING PILOT'S FLIGHT OPERATING INSTRUCTIONS
FLYING WING
RESTRICTED
Reprinted by Periscopefilm.com

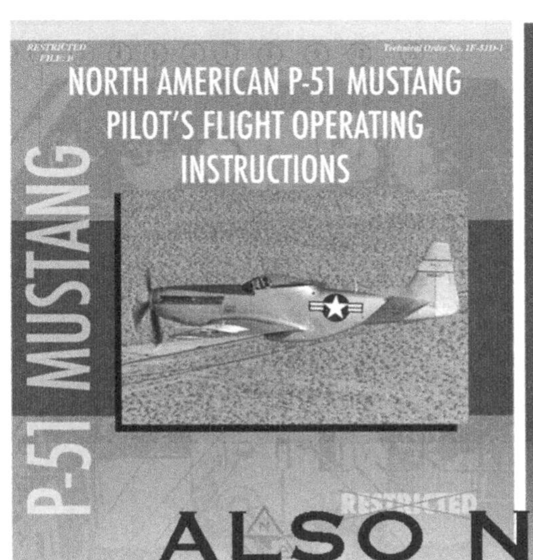
RESTRICTED FILE: B
NORTH AMERICAN P-51 MUSTANG PILOT'S FLIGHT OPERATING INSTRUCTIONS
P-51 MUSTANG
RESTRICTED
Originally Published by USAAF and North American Aviation
Reprinted by Periscopefilm.com

RESTRICTED FILE: B
B-17F AIRPLANE
PILOT'S FLIGHT OPERATING INSTRUCTIONS
B-17 F AIRPLANE
Originally published by the United States Army Air Forces December, 1942.
RESTRICTED
Reprinted by Periscopefilm.com

RESTRICTED FILE: B
LOCKHEED SR-71 BLACKBIRD PILOT'S FLIGHT OPERATING INSTRUCTIONS
SR-71 BLACKBIRD
RESTRICTED
Originally Published by Lockheed and the USAF

ISBN #978-1-4116-8900-8